A First Course in

MATHEMATICAL STATISTICS

A First Course in
MATHEMATICAL STATISTICS

by

C. E. WEATHERBURN
M.A., D.Sc., Hon. LL.D.

*Emeritus Professor of Mathematics in the University
of Western Australia*

CAMBRIDGE

AT THE UNIVERSITY PRESS

1968

Published by the Syndics of the Cambridge University Press
Bentley House, 200 Euston Road, London, N.W.1
American Branch: 32 East 57th Street, New York, N.Y. 10022

Standard Book Number:
521 06750 2 clothbound
521 09158 6 paperback

First Edition	1946
Reprinted	1947
Second Edition	1949
Reprinted	1952
Reprinted with corrections	1957
Reprinted	1961
	1962
	1968

First printed in Great Britain
at the University Printing House, Cambridge
Reprinted by photolithography in Great Britain
by Bookprint Limited, Crawley, Sussex

CONTENTS

Chapter I

FREQUENCY DISTRIBUTIONS

Chapter II

PROBABILITY AND PROBABILITY DISTRIBUTIONS

Chapter III

SOME STANDARD DISTRIBUTIONS

Binomial and Poissonian Distributions

Contents

The Normal Distribution

Chapter IV

BIVARIATE DISTRIBUTIONS. REGRESSION AND CORRELATION

Chapter V

FURTHER CORRELATION THEORY. CURVED REGRESSION LINES

Curved Regression Lines

Chapter VI

THEORY OF SIMPLE SAMPLING

Sampling of Attributes

Sampling of Values of a Variable

Chapter VII

STANDARD ERRORS OF STATISTICS

Sampling from a Bivariate Population

Chapter VIII

BETA AND GAMMA DISTRIBUTIONS

Chapter IX

CHI-SQUARE AND SOME APPLICATIONS

Chapter X

FURTHER TESTS OF SIGNIFICANCE.
SMALL SAMPLES

Chapter XI

ANALYSIS OF VARIANCE AND COVARIANCE

Analysis of Variance

Analysis of Covariance

Chapter XII

MULTIVARIATE DISTRIBUTIONS.
PARTIAL AND MULTIPLE CORRELATIONS

TABLES

PREFACE

The object of this work is to provide a mathematical text on the Theory of Statistics, adapted to the needs of the student with an average mathematical equipment, including an ordinary knowledge of the Integral Calculus. The subject treated in the following pages is best described not as Statistical Methods but as Statistical Mathematics, or *the mathematical foundations of the interpretation of statistical data.* The writer's aim is to explain the underlying principles, and to prove the formulae and the validity of the methods which are the common tools of statisticians. Numerous examples are given to illustrate the use of these formulae; but, in nearly all cases, heavy arithmetic is purposely avoided in the desire to focus the attention on the principles and proofs, rather than on the details of numerical calculation.

The treatment is based on a course of about sixty lectures on Statistical Mathematics, which the author has given annually in the University of Western Australia for several years. This course was undertaken at the request of the heads of some of the science departments, who desired for their students a more mathematical treatment of the subject than those usually provided by courses on Statistical Methods. The class has included graduates and undergraduates whose researches and studies were in Agriculture, Biology, Economics, Psychology, Physics and Chemistry. On account of such a diversity of interest the lectures were designed to provide a mathematical basis, suitable for work in any of the above subjects. No technical knowledge of any particular subject was assumed.

The first five chapters deal with the properties of distributions in general, and of some standard distributions in particular. It is desirable that the student become familiar with these, before being confronted with the theory of sampling, in which he is required to consider two or more distributions simultaneously. However, the reader who wishes to make an earlier start with sampling theory may take Chapters VI and VII (as far as §60) immediately after Chapter III, since these are independent of the theory of Correlation. The theory of Partial and Multiple Correlations has been left for the final chapter, in order not to delay the study of sampling theory and tests of significance. But those who wish to study this subject earlier may read this chapter immediately after Chapter v, or even after Chapter iv. This order, however, is not recommended for the beginner.

A feature of the book is the use of the properties of Beta and Gamma variates in proving the sampling distributions of the statistics, which are the basis of the common tests of significance. Consequently a special chapter (viii) is devoted to the properties of these variates and their distributions. The treatment is simple, and does not assume any previous acquaintance with the Beta and Gamma functions. The author believes that the use of these variates brings both simplicity and cohesion to the theory. The student is strongly urged to master the theorems of Chapter viii before proceeding to tests of significance. In the preparation of this chapter, and the following one, much help was derived from the study of a recent paper by D. T. Sawkins.*

The considerations, which determined the presentation of the subject of Probability in Chapter ii, are the mathematical attainments of the students for whom the book is intended, and their requirements in studying the remaining chapters. The approach decided on is substantially that of the classical theory. After the proof of Bernoulli's theorem, the relation between the *a priori* definition of probability and the statistical (or empirical) definition is considered; and the measure of probability by a relative frequency in each case is emphasized. If the book had been intended primarily for mathematical specialists, a different presentation of the theory would have been given. But it is futile to expect the average research worker to appreciate an exposition like Kolmogoroff's† or Cramér's,‡ based on the theory of completely additive

---

* 'Elementary presentation of the frequency distributions of certain statistical populations associated with the normal population.' *Journ. and Proc. Roy. Soc. N.S.W.* vol. 74, pp. 209-39. By D. T. Sawkins, Reader in Statistics at Sydney University.

† *Grundbegriffe der Wahrscheinlichkeitsrechnung*, Berlin, 1933.

‡ *Random Variables and Probability Distributions*, University Press, Cambridge, 1937.

set functions. The elementary properties of the moment generating function and the cumulative function are also given in Chapter II, and are used throughout the book. These functions are introduced, not as essential concepts, but as useful instruments which lead to simpler proofs of various theorems. What has been written about them should convince the reader that they deserve his careful attention.

I wish to express my appreciation of the care bestowed on this book by the staff of the Cambridge University Press, and my pleasure in the excellence of the printing. My thanks are due to Professor R. A. Fisher and Messrs Oliver and Boyd for permission to print Tables 3, 4, 5 and 7, which are drawn from fuller tables in Fisher's *Statistical Methods for Research Workers*. I am also indebted to Professor G. W. Snedecor and the Iowa Collegiate Press for permission to reproduce Table 6, which is extracted from a more complete table in Snedecor's *Statistical Methods*. Lastly I wish to thank Mr D. T. Sawkins for help received in correspondence concerning statistical theory, and Mr Frank Gamblen for assistance in reading the final proof.

C.E.W.

PERTH, W.A.
April, 1946

NOTE ON THE SECOND EDITION

The call for reprinting has given an opportunity to correct a number of small errors and misprints throughout the book, and to add a new reference here and there. Paragraph 91 on page 195 is new to this edition.

1949

C.E.W.

CHAPTER I

FREQUENCY DISTRIBUTIONS

1. Arithmetic mean. Partition values

Consider a group of N persons in receipt of wages. Let x shillings be the wage of an individual on some specified day. Then, in general, x will be a variable whose value changes with the individual. Possibly the N values of x will not all be different. Suppose there are only n different values x_1, x_2, ..., x_n which occur respectively $f_1, f_2, ..., f_n$ times. The numbers f_i, in which the subscript i takes the positive integral values 1, 2, ..., n, are called the *frequencies* of the values x_i of the variable x; and the assemblage of values x_i, with their associated frequencies, is the *frequency distribution* of wages for that group of persons on the day specified. The sum of the frequencies is clearly equal to the number of persons in the group, so that

$$N = f_1 + f_2 + ... + f_n = \sum_{i=1}^{n} f_i, \qquad (1)$$

where, as the equation indicates, $\sum_{i=1}^{n} f_i$ denotes the sum of the n frequencies f_i, i taking the integral values 1 to n. When the range of values of i is understood, the sum will be denoted simply by $\sum_i f_i$ or $\sum f_i$. The number N is the *total frequency*. The *mean* of the distribution is the arithmetic mean $\bar{x}$ of the N values of the variable, and is therefore given by

$$\bar{x} = \frac{1}{N} (f_1 x_1 + f_2 x_2 + ... + f_n x_n) = \frac{1}{N} \sum_{i=1}^{n} f_i x_i, \qquad (2)$$

since the value x_i occurs f_i times. This formula expresses what is meant by saying that $\bar{x}$ is the *weighted mean* of the different values x_i, whose weights are their frequencies f_i.

Frequency distributions of many different variables will occur in the following pages. Thus the values of the variable x may be the heights, or the weights, or the ages of a group of persons, or the yields of grain per acre from a number of plots of land. For each finite distribution f_i will denote the frequency of the value x_i. The total frequency N is then given by (1), and the mean $\bar{x}$ of the distribution by (2).

Example. The student to whom the above summation notation is new, may profitably verify the following relations. If a is a constant,

$$\sum_i a f_i x_i = a \sum_i f_i x_i,$$

$$\sum_i f_i (x_i + a) = \sum_i f_i x_i + N a,$$

$$\sum_i f_i (x_i + a)^2 = \sum_i f_i x_i^2 + 2 a \sum_i f_i x_i + N a^2.$$

If (x_i, y_i) is a pair of corresponding values of two variables, x and y, with frequency f_i,

$$\sum_i f_i (x_i + y_i) = \sum_i f_i x_i + \sum_i f_i y_i.$$

The symbol used as a subscript in connection with summation is immaterial; but i, j, r, s, t are perhaps most commonly employed.

Suppose that the frequency distribution of x consists of k partial or component distributions, $\bar{x}_j$ being the mean of the jth component and n_j its total frequency, so that

$$N = \sum_{j=1}^{k} n_j.$$

Then that part of the sum $\sum_i f_i x_i$ which belongs to the jth component has the value $n_j \bar{x}_j$, and the relation (2) is equivalent to

$$\bar{x} = \frac{1}{N} \sum_{j=1}^{k} n_j \bar{x}_j. \tag{3}$$

Consequently the mean of the whole distribution is the weighted mean of the means of its components, the weights being the total frequencies in those components.

Again, let u and v be two variables with frequency distributions in which a value of v corresponds to each value of u. Then the values of the variables occur in pairs. Let N be the number of pairs of values (u_i, v_i), and let

$$x_i = u_i + v_i.$$

The arithmetic mean of the N values of x is equal to that of the N values of the second member, which is $\bar{u} + \bar{v}$. Consequently

$$\bar{x} = \bar{u} + \bar{v}, \tag{4}$$

which expresses that the mean of the sum of two variables is equal to the sum of their means; and the result can be extended to the sum of any number of variables. The reader can prove similarly that, if a and b are constants, and

$$x = au + bv,$$

then $$\bar{x} = a\bar{u} + b\bar{v}. \tag{4'}$$

Suppose the N values of the variable in the distribution to be arranged in ascending order of magnitude. Then the *median* is the middle value, if N is odd; while, if N is even, it is the arithmetic mean of the middle pair, or, more generally, it may be regarded as any value in the interval between these middle values. Similarly, the *quartiles*, Q_1, Q_2, Q_3, are those values in the range of the variable which divide the frequency into four equal parts, the second quartile being identical with the median; and the difference between the upper and lower quartiles, $Q_3 - Q_1$, is the *interquartile range*. The *deciles* and *percentiles* are those values which divide the total frequency into ten and one hundred equal parts respectively. The median, quartiles, deciles and percentiles are often spoken of collectively as *partition values*, since each set of values divides the frequency into a number of equal parts. Sometimes they are referred to as *quantiles*.

That value of the variable whose frequency is a maximum is called a *mode*, or *modal value*, of the distribution. When, as usually happens, there is only one mode, the distribution is said to be *unimodal*.

We shall presently consider continuous frequency distributions. But it should be pointed out at once that the variable may be either continuous or discrete. A continuous variable is one which is capable of taking any value between certain limits; for example, the stature of an adult man. A discrete variable is one which can take only certain specified values, usually positive integers; for example, the number of heads in a throw of ten coins, or the number of accidents sustained by a worker exposed to a given risk for a given time. Of course an *observed* frequency distribution can only contain a finite number of values of the variable, and in this sense all observed frequency distributions are discrete. Nevertheless, the distinction between continuous and discrete variables will be found to be of importance when we come to study populations and probability distributions. In the next few sections we shall assume that the values of the variables are discrete.

2. Change of origin and unit

The following graphical representation will be found helpful. Taking the usual x-axis with origin O, we may represent the variable x by the abscissa of the current point P. Then $\bar{x}$ is the abscissa of a

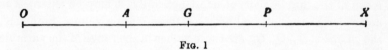

$$O \qquad\qquad A \qquad G \qquad\qquad P \qquad\qquad\qquad X$$

Fig. 1

fixed point G. It is frequently convenient to take a new origin at some point A, whose abscissa is a. Let ξ be the abscissa of P relative to A as origin. Then, since $OP = OA + AP$, we have

$$x = a + \xi.$$

Thus ξ is the excess of x above a, or the *deviation* of x from that value. Taking the mean of each member of this equation we have

$$\bar{x} = a + \bar{\xi}. \tag{5}$$

Thus $\bar{x} - a$, which is the deviation of the mean value of x from a, is equal to $\bar{\xi}$, which is the mean of the deviations of the values x_i from a. In particular, by taking a as $\bar{x}$, we have the result that *the sum of the deviations of the values x_i from their mean is zero*. This is also easily proved directly; for

$$\sum_i f_i(x_i - \bar{x}) = \sum_i f_i x_i - N\bar{x} = 0 \tag{6}$$

in virtue of (2).

In addition to choosing A as origin it may be convenient to use a different unit, say c times the original unit. Then, if u is the deviation of P from A measured in terms of this new unit,

$$u = (x-a)/c$$

or
$$x = a+cu. \tag{7}$$

Taking the mean value of the variable represented by either side we have, in virtue of (4′),

$$\bar{x} = a+c\bar{u}, \tag{8}$$

$\bar{u}$ being the mean value of u for the distribution.

Example. Eight coins were tossed together, and the number x of heads resulting was observed. The operation was performed 256 times; and the frequencies that were obtained for the different values of x are shown in the following table. Calculate the mean, the median and the quartiles of the distribution of x.

x	f	ξ	$f\xi$	$f\xi^2$	$f\xi^3$
0	1	-4	-4	16	-64
1	9	-3	-27	81	-243
2	26	-2	-52	104	-208
3	59	-1	-59	59	-59
4	72	0	0	0	0
5	52	1	52	52	52
6	29	2	58	116	232
7	7	3	21	63	189
8	1	4	4	16	64
Totals	256	—	-7	507	-37

The different values of x are shown in the first column, and their frequencies in the second. The calculation is simplified by taking the value $x = 4$ as origin of ξ. The values of ξ corresponding to those of x are given in the third column, and those of the product $f\xi$ in the fourth. The remaining columns are not needed for the present. Totals for the various columns are given in the bottom row. Hence

$$\bar{\xi} = \frac{1}{N}\sum_i f_i\xi_i = -7/256 = -0\cdot027.$$

The mean value of x is therefore

$$\bar{x} = a+\bar{\xi} = 4-0\cdot027 = 3\cdot973.$$

The mode, being the value of x with the largest frequency, is clearly 4. To find the median we observe that the values of x are arranged in ascending order, and that the 128th and 129th are both 4. Hence the median is also 4. Similarly the 64th and 65th values are both 3, so that the lower quartile is 3. In the same way we find that the upper quartile is 5.

3. Variance. Standard deviation

The *mean square deviation* of the variable x from the value a is, as the name implies, the mean value of the square of the deviation of x from a. It is therefore given by $\frac{1}{N}\sum_i f_i \xi_i^2$. The positive square root of this quantity is the *root-mean-square deviation* from a. In the important case in which the deviation is taken from the mean of the distribution, the mean square deviation is called the *variance* of x, and is denoted by μ_2. The reason for the notation will appear in the next section. The positive square root of the variance is called the *standard deviation* (S.D.) of x, and is denoted by σ. Thus

$$\mu_2 = \sigma^2 = \frac{1}{N}\sum_i f_i (x_i - \bar{x})^2. \tag{9}$$

The variance (or the S.D.) may be taken as an indication of the extent to which the values of x are scattered. This scattering is called *dispersion*. When the values of x cluster closely round the mean, the dispersion is small. When those values, whose deviations from the mean are large, have also relatively large frequencies, the dispersion is large. The concepts of variance and S.D. will play a prominent part in the following pages.

When the mean square deviation from any value a is known, and also the deviation $\bar{\xi}$ of the mean from that value, the variance is easily calculated. For

$$\sigma^2 = \frac{1}{N}\sum_i f_i (\xi_i - \bar{\xi})^2 = \frac{1}{N}\sum_i f_i \xi_i^2 - \frac{2}{N}\bar{\xi}\sum_i f_i \xi_i + \bar{\xi}^2$$

$$= \frac{1}{N}\sum_i f_i \xi_i^2 - \bar{\xi}^2. \tag{10}$$

This formula is of great importance, and will be constantly employed. On multiplying by N we have an equivalent relation, which may be expressed

$$N\sigma^2 = \sum_i f_i \xi_i^2 - \frac{1}{N}(\sum_i f_i \xi_i)^2. \tag{11}$$

Thus $N\sigma^2$ is less than $\sum_i f_i \xi_i^2$, showing that *the sum of the squares of the deviations of the values x_i is least when the deviations are measured from the mean.*

Another possible measure of dispersion is the mean value of the absolute deviation from the mean of the distribution, commonly called the *mean deviation* from the mean. This quantity, however, does not lend itself readily to algebraical treatment, and is therefore not nearly so important as the variance and the S.D. The semi-interquartile range is also sometimes taken as an indication of the magnitude of the dispersion.

The significance of the magnitude of the standard deviation clearly depends upon the values of the variable. Thus a S.D. of 6 in. in the measurements of the height of a tower, is much less significant than an equal S.D. in the measurements of the height of a man. The ratio of the S.D. to the mean value of the variable is called the *coefficient of variation*. It is an absolute measure of dispersion in the sense that it is independent of the unit employed. And by means of this coefficient we are able to compare the variabilities of distributions of different characteristics, such as weight and height. Sometimes the coefficient of variation is defined as 100 times the above value, i.e. as the percentage of the mean which is equal to the S.D.

Example 1. In the example of the preceding section the mean square deviation of x from the value 4 is $507/256 = 1.98$. Hence the variance is given by

$$\sigma^2 = 1.98 - \bar{\xi}^2 = 1.98 - (0.027)^2 = 1.98,$$

and
$$\sigma = 1.407 = 1.41 \text{ nearly.}$$

From the $f\xi$ column it is clear that the sum of the absolute deviations from $x = 4$ is 277. By measuring deviations from the mean, instead of from $x = 4$, we increase the absolute deviations of 161 values, and decrease those of 95 values, by 0.027. Hence the sum of the absolute deviations from the mean is $277 + 66(0.027) = 278.78$. The mean deviation is therefore 1.09 approximately.

Example 2. Find the mean and the variance for the distribution in which the values of x are the positive integers 1, 2, 3, ..., N, the frequency of each being unity.

Here
$$\bar{x} = \frac{N(N+1)}{2N} = \tfrac{1}{2}(N+1).$$

The mean square deviation from $x = 0$ is

$$(1^2 + 2^2 + \ldots + N^2)/N = \tfrac{1}{6}(N+1)(2N+1).$$

Hence
$$\sigma^2 = \tfrac{1}{6}(N+1)(2N+1) - \tfrac{1}{4}(N+1)^2$$
$$= \tfrac{1}{12}(N^2 - 1).$$

Example 3. For the distribution expressed by

$$x = \quad 5 \quad 6 \quad 7 \quad 8 \quad 9 \quad 10 \quad 11 \quad 12 \quad 13 \quad 14 \quad 15$$
$$f = 18 \quad 25 \quad 34 \quad 47 \quad 68 \quad 90 \quad 80 \quad 62 \quad 38 \quad 27 \quad 11$$

the total frequency is 500. Show that the mean value of x is 10·054, the variance 5·58, the s.d. 2·36, the median 10, and the lower and upper quartiles 9 and 12 respectively. Also calculate the mean deviation from the mean as in Ex. 1.

Example 4. A distribution consists of several component distributions. Express the variance of the whole distribution in terms of those of the components and the deviations of the means of the components from the general mean.

Let n_j be the frequency in the jth component, σ_j its s.d., and $d_j = \bar{x}_j - \bar{x}$ the deviation of its mean from the general mean. Then the mean square deviation of this component from the general mean is $\sigma_j^2 + d_j^2$, and the sum of the squares of its deviations from the general mean is $n_j(\sigma_j^2 + d_j^2)$. Hence the variance σ^2 of the whole distribution is given by

$$N\sigma^2 = \sum n_j(\sigma_j^2 + d_j^2),$$

where N is the total frequency $\sum_j n_j$.

4. Moments

In the notation of the preceding sections the mean value of the rth power of the deviation of the variable from the value a is $\sum_i f_i \xi_i^r / N$. This is usually called the rth *moment* of the distribution about the value a, or the moment of order r. The term 'moment' is borrowed from Mechanics. Since f_i/N is the *relative frequency* of the value x_i in the distribution, and the deviation ξ from a is represented by the distance AP, the above expression can be regarded as the sum of the rth moments of the relative frequencies about A. The rth moment about the mean of the distribution is denoted by μ_r. The corresponding moment about a specified value other than the mean, will be denoted* by μ'_r. Thus

$$\mu_r = \frac{1}{N} \sum_i f_i (x_i - \bar{x})^r \tag{12}$$

is the rth moment about the mean, while

$$\mu'_r = \frac{1}{N} \sum_i f_i (x_i - a)^r = \frac{1}{N} \sum_i f_i \xi_i^r \tag{13}$$

* An alternative notation, with ν_r instead of μ'_r, has some advantages.

is the rth moment about the value a. Putting $r = 0$ we see that

$$\mu_0 = \mu_0' = 1. \tag{14}$$

Similarly, in virtue of (2) and (6), we have

$$\mu_1' = \bar{\xi}, \quad \mu_1 = 0. \tag{15}$$

The second moment about the mean is clearly the variance already discussed.

By means of the binomial expansion, moments about the mean of the distribution may be expressed in terms of moments about any other value, $x = a$. Thus

$$\mu_r = \frac{1}{N} \sum_i f_i (\xi_i - \bar{\xi})^r$$
$$= \mu_r' - \binom{r}{1} \bar{\xi} \mu_{r-1}' + \binom{r}{2} \bar{\xi}^2 \mu_{r-2}' - \dots, \tag{16}$$

$\binom{r}{s}$ denoting the binomial coefficient, often written ^rC_s or C_s^r. In particular, in virtue of (14) and (15),

$$\mu_2 = \mu_2' - 2\bar{\xi}^2 + \bar{\xi}^2 = \mu_2' - \bar{\xi}^2 \tag{17}$$

in agreement with (10). Similarly

$$\left. \begin{array}{l} \mu_3 = \mu_3' - 3\bar{\xi}\mu_2' + 2\bar{\xi}^3, \\ \mu_4 = \mu_4' - 4\bar{\xi}\mu_3' + 6\bar{\xi}^2\mu_2' - 3\bar{\xi}^4, \end{array} \right\} \tag{18}$$

and so on.

In calculating moments it is frequently convenient to change the unit. As in § 2, let u be the measure of the deviation from $x = a$, in terms of a unit c times the original unit, so that $\xi = cu$. Then the rth moment of x about a is

$$\mu_r' = \frac{1}{N} \sum_i f_i \xi_i^r = \frac{c^r}{N} \sum_i f_i u_i^r. \tag{19}$$

Thus the rth moment of the variable x is c^r times the corresponding moment of the variable u.

A distribution is said to be *symmetrical* when the frequencies are symmetrically distributed about the mean, that is to say, when values equidistant from the mean have equal frequencies. For example, the distribution expressed by

$$x = 0 \quad 1 \quad 2 \quad 3 \quad 4 \quad 5 \quad 6 \quad 7 \quad 8$$
$$f = 1 \quad 8 \quad 28 \quad 56 \quad 70 \quad 56 \quad 28 \quad 8 \quad 1$$

is symmetrical about its mean $\bar{x} = 4$. In the case of a symmetrical distribution there is the simplification that all the moments of odd order about the mean are equal to zero, since the terms of the sum in (12) cancel in pairs. In the case of an unsymmetrical distribution, the degree of departure from symmetry is called its *skewness*. More than one measure of this property has been proposed. One of the simplest is μ_3/σ^3, while another is half this expression. These are clearly independent of the unit chosen for the variable, and they vanish if the distribution is symmetrical. Another measure of skewness, proposed by Karl Pearson, will be given later.

Example 1. For the distribution of x in the example of § 2, the third moment about $\xi = 0$ is
$$\mu_3' = -37/256 = -0\cdot145.$$

Hence the third moment about the mean is given by

$$\mu_3 = \mu_3' - 3\bar{\xi}\mu_2' + 2\bar{\xi}^3$$
$$= -0\cdot145 - 3(-0\cdot027)(1\cdot98) + 2(-0\cdot027)^3$$
$$= 0\cdot018 \quad \text{nearly.}$$

The skewness, calculated from the formula μ_3/σ^3, is

$$0\cdot018/2\cdot8 = 0\cdot0064,$$

which is very small.

Example 2. For the distribution in § 3, Ex. 3, show that $\mu_3 = -1\cdot92$, and deduce that $\mu_3/\sigma^3 = -0\cdot146$.

5. Grouped distribution

Frequently the number of different values of the variable represented in the distribution is so large that, for convenience in calculating the moments, it becomes necessary to approximate by grouping the values. In such cases the range of variation of x is usually divided into a number of equal intervals. The group of values falling in a given interval constitutes a *class*; and the number of such values is the *class frequency*. The magnitude of an interval is called the *class interval*. For simplicity of calculation the number of intervals chosen will not be too large, preferably not more than 20, or at the most 25; and, in order that the results may be sufficiently accurate, the number must not be too small, preferably not less than 12. The approximation in the calculation of moments consists in regarding each value as equal to the mid-value of the interval in which it falls. We shall later meet formulae,* known as *Sheppard's adjustments*, giving the corrections that may be applied under certain conditions to the approximate values of the moments calculated as above. We merely mention here that, under the conditions referred to, the correction to the mean may be neglected, while the calculated variance should be reduced by $c^2/12$, c being the magnitude of the class interval. The following is an example of the treatment of a distribution by grouping.

Example. Over a period of years, 570 students were examined in Mathematics I at the annual examinations of the University of Western Australia. The marks gained by students ranged from 0 to 99, all being integers. These

Percentages in Mathematics

Interval	Mid-value	f	u	fu	fu^2	fu^3
0 to 4	2	12	−10	−120	1,200	−12,000
5 to 9	7	13	− 9	−117	1,053	− 9,477
10 to 14	12	13	− 8	−104	832	− 6,656
15 to 19	17	14	− 7	− 98	686	− 4,802
20 to 24	22	23	− 6	−138	828	− 4,968
25 to 29	27	23	− 5	−115	575	− 2,875
30 to 34	32	29	− 4	−116	464	− 1,856
35 to 39	37	34	− 3	−102	306	− 918
40 to 44	42	44	− 2	− 88	176	− 352
45 to 49	47	44	− 1	− 44	44	− 44
50 to 54	52	50	0	−1,042		−43,948
55 to 59	57	52	1	52	52	52
60 to 64	62	61	2	122	244	488
65 to 69	67	41	3	123	369	1,107
70 to 74	72	32	4	128	512	2,048
75 to 79	77	27	5	135	675	3,375
80 to 84	82	23	6	138	828	4,968
85 to 89	87	17	7	119	833	5,831
90 to 94	92	13	8	104	832	6,656
95 to 99	97	5	9	45	405	3,645
Totals	—	570	—	966 − 76	10,914	28,170 −15,778

* See § 16.

were grouped in 20 classes, with a class interval of 5, the class frequencies being as shown in the accompanying table. The mid-values of the intervals are 2, 7, 12, ..., 97. The calculation is simplified by taking the class interval as a new unit ($c = 5$), and the value $x = 52$ as a new origin. The deviations u from this origin, measured in class intervals, are shown in the fourth column. The above choice of origin ensures that the larger frequencies are multiplied by the smaller values of u. In the row corresponding to $u = 0$ the entries for fu, fu^2, etc., are all zero, and need not be recorded. The space may therefore be used to record the sums of the negative numbers above this row. The sums of the positive numbers are similarly indicated at the bottom; and the total for each column is given in the last row.

The mean value of u is therefore

$$\bar{u} = \frac{1}{N}\sum_i f_i u_i = -76/570 = -2/15 = -0{\cdot}133,$$

and the mean percentage

$$\bar{x} = a + c\bar{u} = 52 - 0{\cdot}667 = 51{\cdot}333.$$

The mean square deviation of u from $u = 0$ has the value $10{,}914/570 = 19{\cdot}15$; and the variance of u, which may be denoted by σ_u^2, is

$$\sigma_u^2 = 19{\cdot}15 - \bar{u}^2 = 19{\cdot}13.$$

Consequently $\sigma_u = 4{\cdot}374$, and therefore $\sigma_x = 21{\cdot}87$. If, however, Sheppard's adjustment is made, we find

$$\sigma_u^2 = 19{\cdot}047, \quad \sigma_u = 4{\cdot}365, \quad \sigma_x = 21{\cdot}82.$$

The third moment of u about $u = 0$ has the value $-15{,}778/570 = -27{\cdot}68$. Thus, in virtue of (18),

$$\mu_3 = -27{\cdot}68 + 7{\cdot}65 - 0{\cdot}005 = -20{\cdot}03.$$

The skewness of the distribution, calculated from the formula μ_3/σ^3, is $-0{\cdot}24$ nearly. The reader may verify that the mean deviation of x from the mean is about $17{\cdot}7$.

The partition values may be estimated on the assumption that the frequency in any class is evenly distributed among the values in that class. The reader will find in this way that the median is 53, and the lower and upper quartiles 37 and 66 respectively.

6. Continuous distributions

The distributions considered so far are *discrete* distributions, or distributions of discrete values of the variable. A *continuous* distribution is one in which the variable takes every value between certain limits, a and b. The total frequency is therefore infinite; and so is the frequency within any finite interval, α to β, of the range of the variable. We shall confine our attention to continuous distributions, which are such that the *relative frequency* in the infinitesimal interval $x - \frac{1}{2}dx$ to $x + \frac{1}{2}dx$ is expressible as $f(x)\,dx$, where $f(x)$ is a continuous function of x, called the *relative frequency density*. Then the continuous curve

$$y = f(x) \tag{20}$$

is the relative *frequency curve* for the distribution. If this curve is symmetrical about some line $x = c$, the distribution is said to be *symmetrical*. The infinitesimal interval $x - \frac{1}{2}dx$ to $x + \frac{1}{2}dx$, with mid-value x and magnitude dx, may be conveniently referred to as the interval dx. The relative frequency $f(x)\,dx$ in this interval is represented by the area under the curve (20), between the ordinates at the ends of the interval. Hence the relative frequency for the interval α to β is given by the integral $\int_{\alpha}^{\beta} f(x)\,dx$. The sum of all the relative frequencies is unity, so that

$$\int_{a}^{b} f(x)\,dx = 1. \tag{21}$$

As in the case of a discrete distribution, the *moment* of order r about a specified value is the sum of the rth moments of the relative frequencies about that value. Hence the mean $\bar{x}$, which is the first moment about the origin, is given by

$$\bar{x} = \int_{a}^{b} x f(x)\,dx. \tag{22}$$

The moments of order r, about the origin and the mean, are

$$\mu_r' = \int_{a}^{b} x^r f(x)\,dx, \quad \mu_r = \int_{a}^{b} (x - \bar{x})^r f(x)\,dx.$$

And, in virtue of the binomial expansion, it follows as in §4 that the relations (16)–(18) hold for a continuous distribution also. In particular, the variance of the distribution is given by

$$\sigma^2 = \mu_2 = \int_{a}^{b} x^2 f(x)\,dx - \bar{x}^2. \tag{23}$$

Example 1. By way of illustration consider a straight rod of length l. The distance x of a molecule of the rod from one end may be regarded as a continuous variable, ranging from 0 to l; and the distribution of the values of x is then continuous. If M is the mass of the rod, and the linear density m is continuous, the function $f(x)$ for the distribution has the value m/M.

In the case of a uniform rod, of length $2a$, the distance x of a molecule from the middle point varies continuously from $-a$ to a, and the distribution of x is continuous with $f(x) = 1/2a$. The frequency curve is a straight line parallel to the x-axis, and the distribution of x is said to be *rectangular* or *uniform*. The mean value of x is now zero, and the variance is given by

$$\sigma^2 = \int_{-a}^{a} \frac{x^2\, dx}{2a} = \frac{a^2}{3}.$$

All the moments of odd order about the mean are zero. The moment of order $2n$ is easily shown to be $a^{2n}/(2n+1)$.

Example 2. Draw the frequency curve for the symmetrical distribution in which

$$f(x) = \frac{2a}{\pi}\left(\frac{1}{a^2+x^2}\right),$$

the range of the variable being $-a$ to a; and show that $f(x)$ satisfies the condition (21). Also show that the variance is

$$\sigma^2 = a^2(4-\pi)/\pi = 0\cdot273a^2,$$

and the s.d. $\sigma = 0\cdot52a$, while

$$\mu_4 = a^4(1-8/3\pi) = 0\cdot151a^4.$$

It is apparent from the above that a discrete distribution cannot be identified with a continuous one. Sometimes, however, a discrete distribution D_1 approximates to a certain continuous distribution D_2 in the sense that, for any interval in the range of the variable, the relative frequency of D_1 is approximately equal to that of D_2. Such an approximation requires that the total frequency N of D_1 should be large, and that there should be only very small intervals between pairs of adjacent values of the variable. Consider, for example, the distribution of ages of all living people, not more than (say) 50 years old at a specified instant. The variable x ranges from 0 to 50 years; and the frequency in any interval, not less than a few hours, is so large that, since a period of a few hours is very small compared with 50 years, the distribution may be regarded as continuous for all practical purposes. If the necessary data were available, we could calculate a continuous function $f(x)$ to give the relative frequency density to any reasonable degree of accuracy.

The diagram illustrates the frequency curve of an unsymmetrical, unimodal, continuous distribution. The *mode* of the distribution is the abscissa of the maximum ordinate NN'. Since the area under the curve in any interval represents the relative frequency for that

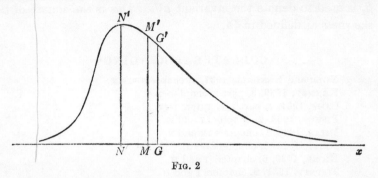

Fig. 2

interval, it follows that the *median* is the abscissa of the ordinate MM', which bisects the area under the curve. And from (22) we see that the *mean* is the abscissa of the ordinate GG', which passes through the centroid of the area under the curve. We may mention in passing that, when the skewness is small, the relation

$$\text{mean} - \text{mode} = 3(\text{mean} - \text{median})$$

holds fairly accurately. The student may regard this as an empirical relation, since we shall not give any proof. The definition of skewness most frequently used is that of Karl Pearson. It is

$$\text{skewness} = \frac{\text{mean} - \text{mode}}{\text{S.D.}}.$$

When the mode can be accurately determined this definition is a convenient one. The frequency curve in the diagram is that of a positively skew distribution, the longer tail of the curve being to the right.

The ratio of the fourth moment about the mean of a distribution, to the square of the variance, is independent of the unit employed. This invariant of the distribution is called its *kurtosis*, and is frequently denoted by β_2. Thus

$$\beta_2 = \mu_4/\mu_2^2. \tag{24}$$

In the normal distribution, to be considered later, the kurtosis has the value 3. Since this distribution is regarded as the standard or ideal, the quantity $\beta_2 - 3$ for any distribution is called its excess of kurtosis, or briefly its *excess*. Corresponding to the above notation, β_1 is used to denote the invariant μ_3^2/μ_2^3. This is the square of the skewness as defined in § 4.

COLLATERAL READING*

YULE and KENDALL, 1937, 1, chapters VI–IX.
KENNEY, 1939, 3, part I, chapters I–V.
CAMP, 1934, 1, part I, chapters I–IV.
FISHER, 1938, 2, chapters I and II.
MILLS, 1938, 1, chapters IV and V.
JONES, 1924, 3, chapters I–VII.
RIDER, 1939, 6, chapters I and II.
TIPPETT, 1931, 2, chapters I and II.
RIETZ (ed.), 1924, 1, chapter II.
GOULDEN, 1939, 2, chapters I and II.
BOWLEY, 1920, 2, part I, chapters V and VI; part II, chapter I.
KENDALL, 1938, 5.

EXAMPLES I

1. A distribution consists of three components with frequencies of 200, 250 and 300, having means of 25, 10 and 15, and standard deviations of 3, 4 and 5 respectively. Show that the mean of the combined distribution is 16, and its S.D. 7·2 approximately.

2. The yields of grain (*x* lb.) from 500 small plots are grouped in classes with a common class interval (0·2 lb.) in the table below, the values of *x* given being the mid-values of the classes. Show that the mean of the distribution is 3·95 lb., its S.D. 0·46 lb., the median also 3·95 lb., and the quartiles 3·63 and 4·28 lb.

x	f	x	f	x	f	x	f	x	f
2·8	4	3·4	47	4·0	88	4·6	35	5·2	4
3·0	15	3·6	63	4·2	69	4·8	10	—	—
3·2	20	3·8	78	4·4	59	5·0	8	—	—

(Mercer and Hall, *Journ. Agric. Sci.* 1911, vol. 4, p. 107.)

* The references are to the literature listed at the end of the book. The student is strongly advised to read some of the literature mentioned at the end of each chapter, preferably in the order given. He will thus acquire a better background for the mathematical theory.

3. The wages of 1,000 employees range from 4s. 6d. to 19s. 6d. They are grouped in 15 classes with a common class interval of 1s., and the class frequencies, from the lowest class to the highest, are 6, 17, 35, 48, 65, 90, 131, 173, 155, 117, 75, 52, 21, 9, 6. Tabulate the data, and show that the mean wage is 12·006s., the S.D. 2·626s. = 2s. 7$\frac{1}{2}d$., and the median 12·127s. = 12s. 1$\frac{1}{2}d$., and the mode 12·369s. = 12s. 4$\frac{1}{2}d$. nearly. (Adjusted S.D. 2·61s.)

4. With the notation of § 4, and by a method similar to that used in proving (16), show that

$$\mu_3' = \mu_3 + 3\bar{\xi}\mu_2 + \bar{\xi}^3,$$
$$\mu_4' = \mu_4 + 4\bar{\xi}\mu_3 + 6\bar{\xi}^2\mu_2 + \bar{\xi}^4,$$
$$\mu_r' = \mu_r + r\bar{\xi}\mu_{r-1} + \binom{r}{2}\bar{\xi}^2\mu_{r-2} + \dots + \binom{r}{2}\bar{\xi}^{r-2}\mu_2 + \bar{\xi}^r.$$

5. The first three moments of a distribution about the value 2 of the variable are 1, 16 and −40. Show that the mean is 3, the variance 15, and $\mu_3 = -86$. Also show that the first three moments about $x = 0$ are 3, 24 and 76.

6. Prove that the mean deviation from the median is less than that measured from any other value. (See Aitken, 1939, I, p. 32.)

7. Show that, if the class interval of a grouped distribution is less than one-third of the calculated S.D., Sheppard's adjustment makes a difference of less than $\frac{1}{2}$ % in the estimate of the S.D.

8. Show that, if the variable takes the values 0, 1, 2, 3, ..., n with frequencies proportional to the binomial coefficients 1, n, $\binom{n}{2}$, $\binom{n}{3}$, ..., n, 1 respectively, then the mean of the distribution is $\frac{1}{2}n$, the second moment about $x = 0$ is $n(n+1)/4$, and the variance is $\frac{1}{4}n$.

9. In a continuous distribution, whose relative frequency density is given by $f(x) = 3x(2-x)/4$, the variable ranges from 0 to 2. Show that the distribution is symmetrical, with mean $x = 1$, and variance 1/5. Show that the second and third moments about $x = 0$ are 6/5 and 8/5 respectively; and verify that $\mu_3 = 0$.

B

10. *Exponential distribution.* Consider the continuous distribution in which $f(x) = a e^{-ax}$, a being positive, and the variable ranging from 0 to ∞. Show that the mean is $1/a$ and the variance $1/a^2$. Also prove that the second and third moments about $x = 0$ are $2/a^2$ and $6/a^3$ respectively, and that $\mu_3 = 2/a^3$.

11. *Factorial moments.* The factorial moments of a distribution are defined as follows. That of order r about the origin $(x = 0)$ is

$$\mu'_{(r)} = \frac{1}{N} \sum_i f_i x_i^{(r)},$$

where

$$x^{(r)} = x(x-1)(x-2) \ldots (x-r+1).$$

Show that factorial moments are related to ordinary moments by the equations

$$\mu'_{(1)} = \mu'_1 = \bar{x}, \quad \mu'_{(2)} = \mu'_2 - \bar{x},$$

$$\mu'_{(3)} = \mu'_3 - 3\mu'_2 + 2\bar{x},$$

$$\mu'_{(4)} = \mu'_4 - 6\mu'_3 + 11\mu'_2 - 6\bar{x},$$

and so on. Similarly

$$\mu'_2 = \mu'_{(2)} + \bar{x}, \quad \mu'_3 = \mu'_{(3)} + 3\mu'_{(2)} + \bar{x},$$

$$\mu'_4 = \mu'_{(4)} + 6\mu'_{(3)} + 7\mu'_{(2)} + \bar{x}.$$

The expression for the factorial moment $\mu_{(r)}$ about the mean is obtained from that defining $\mu'_{(r)}$ by replacing x by the deviation $x - \bar{x}$ from the mean. Relations corresponding to the above are obtained by dropping the dashes and putting $\bar{x} = 0$. The student may then easily verify that factorial moments about the mean are connected with those about the origin by the equations

$$\mu_{(2)} = \mu'_{(2)} - \bar{x}^2 + \bar{x},$$

$$\mu_{(3)} = \mu'_{(3)} - 3\mu'_{(2)}\bar{x} + 2\bar{x}^3 - 2\bar{x},$$

$$\mu_{(4)} = \mu'_{(4)} - 4\bar{x}\mu'_{(3)} + 6\bar{x}(\bar{x}+1)\mu'_{(2)} - 3\bar{x}(\bar{x}^3 + 2\bar{x}^2 - \bar{x} - 2).$$

CHAPTER II

PROBABILITY AND PROBABILITY DISTRIBUTIONS

7. Explanation of terms. Measure of probability

We begin by approaching the subject of probability from the point of view of the classical theory. Later in the chapter we shall give the statistical or empirical approach, and the relation between the two will become apparent. We hope in this way to provide a simple introduction to the subject, adapted to the needs of those for whom the book is written.

The throwing of an ordinary cubical die may result in any one of six different cases, in the sense that any one of the six faces may be uppermost when the die comes to rest. This group of six cases is exhaustive, because it includes all possible cases that may result. The different cases are mutually exclusive, since no two faces can be uppermost at the same time. The throwing of the die may be referred to as a trial. In general a *trial* is the establishing of certain conditions, which must produce one of several results or cases. Two such cases are said to be *mutually exclusive* when the happening of one of them precludes the happening of the other; and a group of cases is said to be *exhaustive* when it includes all possible ones. The number of cases *favourable* to an event A are those that entail the happening of A. For example, in the throwing of the die three of the cases are favourable to the appearance of an even number, and two are favourable to the appearance of a multiple of 3.

Suppose that our die is a perfect cube made of homogeneous material, and that the marking of the faces has not made it dynamically unsymmetrical. Then there is no reason to expect that, as the result of an unbiased throw, any particular face will come uppermost rather than any other. We say that the six cases are equally likely, or equally probable. Similarly the 52 cases that may result from the drawing of a card without discrimination from an ordinary pack, are equally likely. And, in general, two events are said to be *equally* *likely* if, after all relevant evidence has been taken into account, one of them may not be expected rather than the other.*

The terms *probable* and *probability* are used in ordinary language in many different connections; and the reader will agree that, in a great many instances, it is useless to attempt a numerical estimate of the probability under consideration. For instance, we cannot give a numerical value to the probability that a certain man's political or religious beliefs are correct, or that a statement by a perfect stranger is true. There is, however, a very large group of questions in connection with which a numerical estimate of probability may be attempted with very useful results. We shall first state the method of measurement adopted in the classical theory, and then give examples of problems to which the method is applicable.

Measure of probability.† *If a trial may result in any one of n exhaustive, mutually exclusive and equally likely cases, and m of these are favourable to an event A, then the probability (or chance) that A will happen as the result of the trial is measured by the quotient m/n.*

This measure of the probability of the event A is denoted by p. Thus

$$p = \frac{m}{n},\qquad(1)$$

and p is clearly a positive number, not greater than unity. The *opposite event* is understood as the failure of A to happen; and its probability is denoted by q. Since the number of cases involving the failure of A is $n-m$, it follows that

$$q = (n-m)/n = 1-m/n = 1-p,$$

so that $$p+q = 1.\qquad(2)$$

This method of measuring probability is confined to problems in which the results of a trial are reducible to a certain number of equally likely cases. Considerations of symmetry and similarity frequently enable us to decide whether, in the problem before us, the resulting cases are of this nature; and, only if they are so, is the calculation valid. Objection is often raised to the use of the idea of

---

* Cf. Uspensky, 1937, 2, p. 5. † *Ibid.* p. 6.

equally probable cases in defining the measure of probability. The objection loses some of its force if we remember that, in the corresponding concept of temperature, there are methods of recognizing equality of temperature which are quite independent of the measurement either of absolute temperature or of difference in temperature.

When we speak of choosing one *at random* from a group of n objects, we mean that the choice is made in such a way that each object has the same chance of being selected. Various methods have been devised for making such a choice; and these do not assume any similarity on the part of the objects in the group. One such method is illustrated by the procedure often adopted in the case of a lottery. The objects are numbered 1, 2, ..., n, and n similar marbles are numbered correspondingly. The marbles are placed in an urn, and thoroughly mixed; and one of them is then chosen without discrimination. The number on this marble determines the object selected. The method thus uses the similarity of the marbles to ensure that the selection is random. Statisticians have devised other methods of random selection; but it is unnecessary for us to examine them.

Example 1. The chance of throwing a 6 with an ordinary die is 1/6. The chance of throwing an odd number is $3/6 = 1/2$.

Example 2. In a class of 12 pupils, 5 are boys and the remainder girls. The probability that a pupil selected at random will be a girl is 7/12. The probability that two pupils selected at random will both be girls is $\binom{7}{2} \div \binom{12}{2} = \frac{7}{22}$. The *odds against* their being both girls are $15:7$.

Example 3. From each of three married couples one of the partners is selected at random. What is the probability of their being all of one sex?

The number of favourable cases is two, viz. all men or all women. The total number of equally likely cases is $2^3 = 8$. Hence the required probability is 1/4. The odds against are therefore $3:1$.

The probability of choosing two men and one woman is 3/8.

8. Theorems of total and compound probability

The determination of probabilities by direct enumeration of the number of cases is often laborious. The calculation may be simplified by using the theorems of addition and multiplication of probabilities, which are also known as the theorems of total and compound probability. Let us first consider addition.

Suppose that a trial may result in any one of n equally likely cases, of which m_1 are favourable to an event A_1, m_2 to an event A_2, ..., m_k to an event A_k, and so on. If the events A_1, ..., A_k are mutually exclusive, the corresponding favourable cases are all different. The number of cases favourable to either A_1 or A_2, ... or A_k is therefore $m_1 + m_2 + \dots + m_k$. Hence the probability that one of these events will happen is

$$\frac{m_1 + m_2 + \dots + m_k}{n} = \frac{m_1}{n} + \frac{m_2}{n} + \dots + \frac{m_k}{n} = \sum_{i=1}^{k} p_i,$$

where p_i is the probability of the event A_i. We may therefore state the theorem of addition of probabilities:

The probability that one of several mutually exclusive events will happen is the sum of the probabilities of the separate events.

In particular, if the k events A_i are exhaustive, so that they include all the mutually exclusive cases that may arise, the sum of the k numbers m_i is equal to n, and therefore $\sum p_i = 1$. Thus the sum of the probabilities of the exhaustive and mutually exclusive cases that may result from the trial is equal to unity. From the above argument it is also clear that, if an event may occur in several mutually exclusive forms, the probability of the events is the sum of the probabilities of its mutually exclusive forms.

Example 1. What is the probability of obtaining a total of 9 points in a single throw with two dice?

The event may happen in one of the four mutually exclusive forms, 6 and 3, 5 and 4, 4 and 5, 3 and 6, the chance of each of these being 1/36. Hence the required probability is 1/9.

Show that the probability of a total of 7 points is 1/6, and that of a total of 10 points is 1/12.

Example 2. From a set of 17 cards, numbered 1, 2, ..., 17, one is drawn at random. What is the chance that its number is a multiple of 3 or of 7?

The chance of its being a multiple of 3 is 5/17, and that of a multiple of 7 is 2/17. The required probability is therefore 7/17, the two events being mutually exclusive.

Show that the chance that the number will be a multiple of 3 or of 5, or of both, is also 7/17.

Consider next a pair of related trials. Suppose, for instance, that we have an urn containing r red and s white balls, and that we make a random drawing of two balls in succession. The first trial is the drawing of the first ball; and, if this proves to be red, we shall say that the event A has happened. Similarly, the second trial is the drawing of the second ball, without replacing the first; and, if this proves to be red, we shall say that the event B has happened. The probability of A is clearly $r/(r+s)$. That of B depends on whether A has happened or not. Thus, if A has happened at the first trial, the probability of B in the second is $(r-1)/(r+s-1)$; but, if A has not happened, it is $r/(r+s-1)$. The former of these is the *conditional probability* of B on the assumption that A has happened. Thus the two events, A and B, are not independent. Two events are said to be *independent* only when the probability of either of them is not affected by the happening or failure of the other. By way of illustration consider a combined trial consisting of the throwing of two ordinary dice, either together or in succession. If the event A is the throwing of a 6 with the first die, and the event B the throwing of a 5 with the other, the probability of B is 1/6, whether A has happened or not. Events A and B are thus independent.

More generally, suppose that a trial or a combination of trials may result in any one of n equally likely cases, some of which entail the happening of A alone, some that of B alone, and others that of both A and B. Let m be the number favourable to the event A. The cases favourable to both A and B are all included in these m. Let m_1 be their number. Then the probability p that both A and B will happen is given by

$$p = \frac{m_1}{n} = \frac{m}{n}\frac{m_1}{m}.$$

The first quotient, m/n, is the probability of the event A. Also, since m is the number of cases that entail A, and m_1 is the number of these that entail B also, m_1/m is the conditional probability of B on the assumption that A has happened. Hence we have the theorem of compound probability:

The probability of the combined occurrence of two events, A and B, is the product of the probability of A by the conditional probability of B on the assumption that A has happened.

In the case of independence the result may be stated simply, that the probability that two independent events will both happen is the product of the probabilities of the separate events. The theorem may clearly be extended to include any number of events.

Example 3. A coin is tossed three times. Find the chance that head and tail will show alternately.

The required event has two mutually exclusive forms, viz. head-tail-head and tail-head-tail. By the above theorem the chance of either of these is $(\frac{1}{2})^3$; and by the theorem of total probability the required chance is then $(\frac{1}{2})^3 + (\frac{1}{2})^3 = \frac{1}{4}$.

Example 4. Three groups of children contain respectively 3 girls and 1 boy, 2 girls and 2 boys, 1 girl and 3 boys. One child is selected at random from each group. Find the chance that the three selected comprise 1 girl and 2 boys.

The event may happen in any of the mutually exclusive ways girl-boy-boy, boy-girl-boy, boy-boy-girl. The probabilities of these three are $\frac{3}{4} \cdot \frac{1}{2} \cdot \frac{3}{4}$, $\frac{1}{4} \cdot \frac{1}{2} \cdot \frac{3}{4}$, $\frac{1}{4} \cdot \frac{1}{2} \cdot \frac{1}{4}$ respectively. The required probability is their sum, which is 13/32.

9. Probability distributions. Expected value

Suppose that, corresponding to the n exhaustive and mutually exclusive cases that may result from a trial, a variable x assumes the n values x_i with corresponding probabilities p_i. Then the assemblage of values x_i, with their probabilities p_i, constitutes the *probability distribution* of the variable for that trial. A variable, which possesses a probability distribution, is often called a *variate*. Most of the concepts introduced in connection with frequency distributions are equally applicable to probability distributions. Thus the rth moment, μ'_r, of the distribution about the value $x = 0$ is defined by the equation

$$\mu'_r = \sum_i p_i x_i^r, \tag{3}$$

probability taking the place of relative frequency since, as proved in § 8, $\sum p_i = 1$. In particular, the first moment about $x = 0$ is

$$\mu'_1 = \sum_i p_i x_i. \tag{4}$$

Just as in the case of frequency this moment is called the *mean value* of the variate, or of the distribution. More commonly, however, it is referred to as the *expected value* or *expectation* of the variate. It is denoted by $E(x)$, or sometimes by $\bar{x}$, though the former is preferable. Thus

$$E(x) = \sum_i p_i x_i. \tag{5}$$

Let $\psi(x)$ be a function of the variate x. This assumes the value $\psi(x_i)$ when x assumes the value x_i; and p_i is therefore the probability of this value of the function. The expected value of the function is thus given by

$$E[\psi(x)] = \sum_i p_i \psi(x_i).\tag{6}$$

In particular (3) expresses that μ_r' is the expected value of the rth power of the variate.

The rth moment about the mean of the probability distribution is defined by a formula corresponding to § 4 (12). Thus

$$\mu_r = \sum_i p_i(x_i - \bar{x})^r = E([x - E(x)]^r).\tag{7}$$

In particular, the second moment about the mean is the *variance* of the distribution; and its positive square root is the standard deviation, σ. The relation

$$\mu_2 = \mu_2' - \bar{x}^2\tag{8}$$

holds as before. It may be expressed in the alternative notation

$$E([x - E(x)]^2) = E(x^2) - [E(x)]^2.\tag{8'}$$

The relations between μ_r and μ_r' are the same as for a frequency distribution; while, corresponding to § 2 (6), we have the identity

$$E[x - E(x)] = 0.\tag{9}$$

In words, *the expected value of the deviation of a variate from its mean is zero*.

Example 1. What is the expected value of the number of points that will be obtained in a single throw with an ordinary die?

Here the variate is the number of points showing. It assumes the values 1, 2, ..., 6 with probability 1/6 in each case. Hence

$$E(x) = (1 + 2 + \ldots + 6)/6 = 7/2 = 3 \cdot 5.$$

Example 2. From an urn, containing 3 red balls and 2 white, a man is to draw two balls at random without replacement, being promised 20s. for each red ball he draws, and 10s. for each white one. Find his expectation.

For the possible results of the drawing there are three exhaustive and mutually exclusive cases, viz. 2 red balls, 1 red and 1 white, and 2 white. The corresponding probabilities are easily shown to be 3/10, 3/5 and 1/10.

The variable is the amount to be given to the man on the result of the draw; and this has the value 40s. in the first case, 30s in the second, and 20s. in the third. The expected value of the amount to be given is therefore

$$(\tfrac{3}{10} \times 40 + \tfrac{3}{5} \times 30 + \tfrac{1}{10} \times 20)s. = 32s.$$

Example 3. Show that, if c is constant,

$$E(cx) = cE(x), \quad E[c\psi(x)] = cE[\psi(x)].$$

10. Expected value of a sum or a product of two variates

Let x and y be two variates, the first of which assumes the m values x_i with probabilities p_i ($i = 1, 2, ..., m$), and the second assumes the n values y_j with probabilities p'_j ($j = 1, 2, ..., n$). The sum $x + y$ can assume the mn values of $x_i + y_j$, since any of the m values of i may be associated with any of the n values of j. Let p_{ij} denote the probability that x assumes the value x_i and, at the same time, y assumes the value y_j. Then by (5)

$$
\begin{aligned}
E(x+y) &= \sum_{i=1}^{m} \sum_{j=1}^{n} p_{ij}(x_i + y_j) \\
&= \sum_i \sum_j p_{ij} x_i + \sum_i \sum_j p_{ij} y_j \\
&= \sum_i p_i x_i + \sum_j p'_j y_j.
\end{aligned}
$$

For $\sum_j p_{ij}$, being the sum of the probabilities that x assumes the value x_i while y assumes one of the values $y_1, y_2, ..., y_n$, is equal to p_i. Similarly $\sum_i p_{ij} = p'_j$. Consequently

$$E(x+y) = E(x) + E(y). \tag{10}$$

Thus *the expected value of the sum of two variates is equal to the sum of their expected values*; and the theorem may be extended to the sum of any number of variates.

The variates, x and y, are said to be *independent** if the probability

---

* Or, *statistically independent*. Since this is the kind of independence with which we are chiefly concerned in these pages, the adverb 'statistically' will usually be omitted. Statistical independence has been described by Aitken (1939, 1, p. 148) as 'obedience to the multiplication theorem of probability'. This will be apparent to the student after he read §§ 12 and 31 below. Statistical independence should not be confused with *functional* independence. The variables x and y are functionally dependent when there exists a functional relation $F(x, y) = 0$ which holds identically. They are functionally independent if no such relation exists.

that either of them will assume a prescribed value does not depend on the value assumed by the other. We proceed to examine the expected value of the product, xy, of two independent variates. In the above notation this product may assume the mn mutually exclusive values $x_i y_j$. The probability that the product will assume a particular value, $x_i y_j$, is $p_i p'_j$, being the product of the probabilities that x will assume the value x_i and y the value y_j. Hence the expected value of the product is given by

$$E(xy) = \sum_{i=1}^{m} \sum_{j=1}^{n} p_i p'_j x_i y_j.$$

Performing first the summation with respect to j, and then that with respect to i, we have

$$E(xy) = \sum_i p_i x_i E(y) = E(y) \sum_i p_i x_i$$

$$= E(x) \cdot E(y). \tag{11}$$

Thus *the expected value of the product of two independent variates is equal to the product of their expected values.*

If the mean of either of the independent variates, say x, is taken as origin for that variate, then $E(x) = 0$, and consequently $E(xy) = 0$. In particular this relation holds when both the variates are measured from their means. The expected value of the product of the deviations of the two variates from their means is called their *covariance*. Thus *the covariance of two independent variates is equal to zero.* From this we may deduce the important theorem:

The variance of the sum of two independent variates is equal to the sum of their variances.

If the two variates are measured from their means the variance of their sum, being the expected value of $(x+y)^2$, is given by

$$E(x^2 + 2xy + y^2) = E(x^2) + E(y^2),$$

and is therefore equal to the sum of the variances of x and y. The theorem may clearly be extended to the sum of any number of variates which are independent in pairs.

11. Repeated trials. Binomial distribution

Suppose that a trial is repeated, so that we have a series of n trials. The happening of the event A as the result of a trial will be called a success. We consider first the case in which the trials are independent, and the probability p of success is the same for each trial. Then the probability of failure is q, where $p+q = 1$. The probability that there will be exactly r successes in the series of n trials, and therefore $n-r$ failures, is easily found. For, by the theorem of compound probability, the chance of r successes and $n-r$ failures in a specified order is $p^r q^{n-r}$. But the number of different orders in which these successes and failures may occur is $\binom{n}{r}$, being the number of ways of selecting r out of the n positions for the successes. Consequently, by the theorem of total probability, the chance P of exactly r successes in the series of trials is given by

$$P = \binom{n}{r} p^r q^{n-r}. \tag{12}$$

Thus the probabilities of 0, 1, 2, ..., n successes, in a series of n trials for an event of constant probability p, are the respective terms of the binomial expansion

$$(q+p)^n = q^n + nq^{n-1}p + \ldots + nqp^{n-1} + p^n.$$

The probability distribution of the number of successes thus determined is called the *binomial distribution*. Thus the binomial distribution is that in which the variate assumes the values 0, 1, 2, ..., n with corresponding probabilities q^n, $nq^{n-1}p$, ..., p^n.

We may show that *the expected value of the variate in the binomial distribution* * *is np, and its variance npq*. In terms of repeated trials, the expected value of the number of successes in one trial is p, since the variate assumes the values 1 and 0 with probabilities p and q respectively. And, since the number of successes in n trials is the sum of the numbers of successes in the individual trials, it follows from (10) that the expected value of this number is np. Similarly, to find the variance of the distribution we observe that,

* Other proofs of these properties will be given in § 15, Ex. 3, and § 17.

in one trial, the square of the number of successes takes the values
1 and 0 with probabilities p and q respectively, and the expected
value of x^2 in one trial is therefore p. Hence the variance of the
number of successes in one trial is

$$E(x^2) - [E(x)]^2 = p - p^2 = pq.$$

And since the number of successes in n trials is the sum of the numbers in the individual trials, and these trials are independent, the
variance of the total number of successes is the sum of the variances
for the separate trials, and is therefore npq. The s.d. of the number
of successes in n trials is $\sqrt{(npq)}$; and the s.d. of the proportion of
successes in the series is therefore $\sqrt{(pq/n)}$.

Example 1. Find the *most probable* number of successes in the above series
of trials, that is to say, the number of successes which has a greater probability than any other.

The chance of $r+1$ successes will be greater than that of r if

$$\frac{n-r}{r+1}\frac{p}{q} > 1,$$

that is, if $(n+1)p > r+1.$

Hence the most probable number of successes is the integral part of $(n+1)p$.
If, however, $(n+1)p$ is an integer, $r+1$, the chance of $r+1$ successes is equal
to that of r successes, and is greater than that of any other number. Thus, if
$p = 2/5$, the most probable number of successes in 20 trials is the integral
part of $21 \times 2/5$, which is 8. In 24 trials, however, 9 and 10 are the most
probable numbers.

Example 2. A's chance of winning a game against B is $2/3$. Find his
chance of winning at least three games out of five.

This chance is the sum of the probabilities that A will win 3, 4, and 5 games,
and is therefore

$$\binom{5}{3}\left(\frac{2}{3}\right)^3\left(\frac{1}{3}\right)^2 + \binom{5}{4}\left(\frac{2}{3}\right)^4\left(\frac{1}{3}\right) + \binom{5}{5}\left(\frac{2}{3}\right)^5 = \frac{192}{243}.$$

Example 3. *Poisson's series of trials* is a series of n trials in which the
probabilities of success are $p_1, p_2, p_3, ..., p_n$ respectively. Show that the
expected value of the number of successes is $\sum\limits_{i=1}^{n} p_i$, and the variance $\sum\limits_{i} p_i q_i$.

By the same argument as above, the expected value of the number of
successes at the ith trial is p_i, and its variance $p_i q_i$. Hence the result.

Example 4. Verify from (7) that a distribution, in which the variate takes
the values 1 and 0 with probabilities p and q respectively, has variance pq.

12. Continuous probability distributions

As in the case of frequency, a continuous distribution is one in which the variable may take any value between certain limits, a and b. The number of different values is then infinite, and it is useless to speak of the probability of any particular value. Instead of this we consider the probability that the value of the variate will fall within a specified interval in the range of the variate. We confine our attention to cases in which the probability that the value of the variate will fall within the infinitesimal interval $x - \frac{1}{2}dx$ to $x + \frac{1}{2}dx$ is expressible in the form $\phi(x)\,dx$, where $\phi(x)$ is a continuous function of x, called the *probability density* or the probability function. It is, of course, never negative. The continuous curve $y = \phi(x)$ is called the *probability curve*; and, when this is symmetrical, the distribution is said to be *symmetrical*. The area under the curve from $x = \alpha$ to $x = \beta$ represents the probability that the value of x will fall within the interval α to β. The total area under the curve is unity. Thus

$$\int_a^b \phi(x)\,dx = 1. \tag{13}$$

When $\phi(x)$ is constant, the variate is said to have a *uniform* or *rectangular* distribution of probability. The value of the constant must be $1/(b-a)$, in virtue of (13).

The rth moment of the distribution about a particular value is the rth moment of the probability about that value. Since $\phi(x)\,dx$ is the probability corresponding to the interval dx, the rth moment about $x = 0$ is given by

$$\mu'_r = \int_a^b x^r \phi(x)\,dx. \tag{14}$$

In particular, the expected value of x, being the first moment about $x = 0$, is

$$E(x) = \mu'_1 = \int_a^b x\phi(x)\,dx, \tag{15}$$

and the expected value of the function $\psi(x)$ is

$$E[\psi(x)] = \int_a^b \psi(x)\,\phi(x)\,dx. \tag{16}$$

The rth moment about the mean of the distribution is

$$\mu_r = \int_a^b [x - E(x)]^r \phi(x)\, dx, \tag{17}$$

which is the expected value of the rth power of the deviation of x from the mean. The variance of x, being the second moment about the mean, is given by

$$\mu_2 = \int_a^b [x - E(x)]^2 \phi(x)\, dx$$

or $$\sigma^2 = \int_a^b x^2 \phi(x)\, dx - [E(x)]^2. \tag{18}$$

The theorems of § 10 are equally true for continuous distributions. Let the values of x range from a to b, and those of a second variate y from c to d. We may assume that the probability that x falls in the interval dx and, at the same time, y falls in the interval dy is jointly proportional to dx and dy, and expressible in the form $\phi(x, y)\, dx\, dy$, where $\phi(x, y)$ is a continuous function of the two variates. Then the expected value of the sum of the variates is

$$E(x + y) = \int_a^b \int_c^d (x + y)\, \phi(x, y)\, dx\, dy$$
$$= \int_a^b \int_c^d x \phi(x, y)\, dx\, dy + \int_a^b \int_c^d y \phi(x, y)\, dx\, dy.$$

In the first of these two integrals let the integration with respect to y be performed first. Then $\int_c^d \phi(x, y)\, dx\, dy$ is the probability that x will fall in the interval dx irrespective of the value of y. Denote it by $\phi_1(x)\, dx$. Similarly, in the second integral, $\int_a^b \phi(x, y)\, dx\, dy$ is the probability that y will fall in the interval dy irrespective of the value of x. Denote it by $\phi_2(y)\, dy$. Then

$$E(x + y) = \int_a^b x \phi_1(x)\, dx + \int_c^d y \phi_2(y)\, dy$$
$$= E(x) + E(y), \tag{19}$$

as required.

In considering the expected value of the product xy we assume that the variates are independent. Then the probability that the value of x falls in the interval dx is independent of the value of y, and may be expressed as $\phi_1(x)\,dx$. Similarly, the probability that y falls in the interval dy is independent of the value of x, and is expressible in the form $\phi_2(y)\,dy$. Therefore the probability that x falls in the interval dx and, at the same time, y falls in the interval dy is $\phi_1(x)\phi_2(y)\,dx\,dy$. Accordingly, the expected value of the product xy is

$$E(xy) = \int_a^b \int_c^d xy\phi_1(x)\phi_2(y)\,dx\,dy$$
$$= \int_a^b x\phi_1(x)\,dx \int_c^d y\phi_2(y)\,dy = E(x)\,.\,E(y). \qquad (20)$$

Thus the expected value of the product of two independent variates is equal to the product of their expected values. And it follows, as in § 10, that the covariance of two independent variates is equal to zero.

Example 1. Show that, if the variable x has uniform distribution of probability over the range $-a$ to $+a$, then $\phi(x) = 1/2a$, $\bar{x} = 0$, $\sigma^2 = \frac{1}{3}a^2$, the moments of odd order about $x = 0$ are zero, and that of order $2n$ is $a^{2n}/(2n+1)$.

Example 2. Through a point B on the y-axis, whose ordinate is positive and equal to a, a straight line is drawn in a direction taken at random in the interval $\theta = -\frac{1}{4}\pi$ to $\theta = \frac{1}{4}\pi$, θ being the inclination of the line to BO. Examine the probability distribution of the intercept x on the x-axis.

We interpret the data as meaning that the variable θ has uniform distribution of probability in the interval $-\frac{1}{4}\pi$ to $\frac{1}{4}\pi$. Hence the probability that θ will fall in the interval $d\theta$ is $2d\theta/\pi$. Now the intercept on the x-axis has the value $x = a\tan\theta$, so that $\theta = \arctan x/a$, and therefore

$$d\theta = a\,dx/(a^2 + x^2).$$

But, when θ falls in the interval $d\theta$, x falls in the corresponding interval dx. Hence the probability density for the distribution of x is

$$\phi(x) = \frac{2a}{\pi(a^2 + x^2)}.$$

This is also the relative frequency density of the distribution in § 6, Ex. 2, and the properties of the distributions are the same. There is symmetry about $x = 0$, and the s.d. is $0.52a$.

13. Theorems of Tchebychef and Bernoulli

Consider first a theorem due to Tchebychef. Let x be a variate, either discrete or continuous, with s.d. σ. The theorem to be proved fixes an upper limit to the probability that a value of the variate, chosen at random, will differ from the mean by more than $\lambda\sigma$, where λ is a given positive number. *Tchebychef's theorem* may be stated:

In a random choice of a value of a variate, whose standard deviation is σ, the probability that the value chosen will differ from the mean by more than $\lambda\sigma$ does not exceed $1/\lambda^2$.

To prove it we observe that the variance σ^2 is the second moment of the probability of the whole distribution about the mean. Now, if the combined probability P of values further than $\lambda\sigma$ from the mean were greater than $1/\lambda^2$, the second moment of the probability of these values alone would exceed $(\lambda\sigma)^2/\lambda^2$, i.e. σ^2, and that of the whole distribution would, *a fortiori*, be greater than σ^2. Since this is not so the statement in the theorem must be true. Thus

$$P \leqslant \frac{1}{\lambda^2}. \tag{21}$$

In particular the probability that a value of the variate will differ from the mean by more than 3σ does not exceed $1/9$. The theorem is a very conservative one, since the actual value of P is usually very much less than $1/\lambda^2$. The result, however, applies to all distributions; and we shall now use it to prove a theorem due to Bernoulli.

Let m be the number of successes obtained in n independent trials, in which the constant probability of the occurrence of the event A is p. The quotient m/n is the relative frequency of successes. How does this quotient behave as n increases indefinitely? It is a matter of common knowledge that, in trials of this nature in which the value of p is known, the relative frequency obtained is usually a close approximation to p when n is large. But there is no proof, based on the measure of probability given in § 7, that the relative frequency of successes will have p as limiting value when n tends to infinity. James Bernoulli, however, proved that, given any positive number ϵ however small, the probability of $|m/n - p|$ exceeding ϵ tends to zero as n tends to infinity. This is expressed in modern terminology by saying that *m/n converges in probability to p as n* tends to infinity. *Bernoulli's theorem* may also be expressed in the form:

> *Let ε and η be two given positive numbers, however small, and let m be the number of successes in n independent trials, in which the constant probability of success is p. Then the probability that the inequality*

$$\left|\frac{m}{n}-p\right| < \epsilon \tag{22}$$

> *will hold is greater than* $1 - \eta$, *provided that n is greater than a certain number N, depending on ε and η.*

A simple proof may be given as follows. It was shown in § 11 that the mean value of the relative frequency of successes in such a series of trials is p, and its variance σ^2 is pq/n. If then we write $\epsilon = \lambda\sigma$, the condition

$$\left|\frac{m}{n}-p\right| > \epsilon$$

is the condition that the relative frequency of successes should differ from its mean by more than $\lambda\sigma$. But, by Tchebychef's theorem, the probability of this does not exceed $1/\lambda^2$, so that

$$P \leqslant \frac{\sigma^2}{\epsilon^2} = \frac{pq}{n\epsilon^2},$$

and this is less than η for all values of n greater than $pq/\eta\epsilon^2$. Hence Bernoulli's theorem.

The theorem does not assert that the inequality (22) must hold for all values of n greater than N, but that the probability of its not holding is less than η. Even this probability, however small, leaves room for the possibility that it may not hold on some particular occasion. Bernoulli's theorem explains the practice of taking m/n, for a large value of n, as an approximation to the value of p. Indeed, it frequently happens that this use of relative frequency is our only means of estimating the probability of the event.

14. Empirical definition of probability

As already indicated the definition of probability in terms of equally likely cases does not lend itself to every instance in which

a numerical evaluation of probability is desired. Another definition
of the probability of an event is sometimes given in terms of the
relative frequency of the occurrence of the event in an extended
series of trials. The fundamental assumption for such a definition
is that this relative frequency, in a uniform series of trials, tends to
a definite limit as the number of trials in the series tends to infinity;
and this limit is taken as the measure of the probability of the
occurrence of the event in another such trial. No proof can be given
that the above relative frequency does tend to a limit. Convergence
in probability, mentioned in connection with Bernoulli's theorem,
is a consequence of the original definition, and is not the same
thing as the convergence to a limit assumed in the empirical
definition. But, though the two approaches are not theoretically
equivalent, they may be regarded as in agreement for practical
purposes.

By means of the assumption on which the empirical definition is
based, the laws of addition and multiplication of probabilities can
be deduced. Suppose, as in § 8, that a trial may result in any one of
the mutually exclusive events A_i. In a series of n trials let m_i be
the number of times in which the event A_i happens. Then the probability p_i of the happening of this event is given by

$$p_i = \lim_{n \to \infty} \frac{m_i}{n}.$$

Since the number of times in which one of the events $A_1, A_2, ..., A_k$
happens is $\sum_{i=1}^{k} m_i$, the probability of the happening of one of these
events is

$$p = \lim \sum_{i=1}^{k} \frac{m_i}{n} = \sum_{i=1}^{k} \lim \frac{m_i}{n} = \sum_{i=1}^{k} p_i,$$

and we have the theorem of total probability as in § 8.

Suppose next that A and B are different events that may happen
as the results of specified trials T and T' respectively. We require
the probability that, in a pair of such trials, both events will happen.
In a series of n pairs of trials let m be the number of times in which A
happens, and m_1 the number of times in which both A and B happen.

These m_1 occasions are all included in the m occasions on which A happens. Then the probability of the happening of both events is

$$p = \lim_{n \to \infty} \frac{m_1}{n} = \lim_{n \to \infty} \left(\frac{m}{n} \cdot \frac{m_1}{m} \right) = \left(\lim_{n \to \infty} \frac{m}{n} \right) \left(\lim_{m \to \infty} \frac{m_1}{m} \right),$$

since m tends to infinity with n, if the probability of A is not zero. Now $\lim m/n$ is the probability of A, and $\lim m_1/m$ is the conditional probability of B on the assumption that A has happened. We thus have the theorem of compound probability as in § 8. In the particular case of independence the probability of the occurrence of the two events is simply the product of the probabilities of the separate events. As we have already seen, the theorems of total and compound probability are the foundations of the mathematical theory. The measure of probability by relative frequency is also fundamental. In the *a priori* definition it is the relative frequency of favourable cases in the total number of cases, while, in the empirical definition, it is the limit of the relative frequency of the happenings of the event under consideration.

15. Moment generating function and characteristic function

Let $\phi(x)$ be the probability density in the distribution of the variate x. The expected value of e^{tx} is a function of t given by

$$M(t) = \int e^{tx} \phi(x) \, dx, \qquad (23)$$

where the integration is taken over the whole range of x. When this integral has a meaning for a certain range of values of t, we may expand the exponential and integrate term by term, thus obtaining the formula

$$M(t) = 1 + \mu_1' t + \mu_2' t^2/2! + \mu_3' t^3/3! + ..., \qquad (24)$$

where $\qquad \mu_r' = \int x^r \phi(x) \, dx,$

being the moment of order r about the origin $(x = 0)$. For this reason the function $M(t)$, defined by (23), is called the *moment generating function* (m.g.f.) of the distribution about the value $x = 0$. Similarly the m.g.f. about the value $x = a$ is defined as the expected value of $\exp[t(x-a)]$. Denoting this by $M_a(t)$

we have then

$$M_a(t) = \int \exp\left[t(x-a)\right] \phi(x)\, dx, \qquad (25)$$

provided the integral has a meaning; and the expansion of the exponential, and term by term integration, then show that the coefficient of $t^r/r!$ is the moment of order r about the value $x = a$. Since the factor e^{-at} is independent of x, it follows from (25) that

$$M_a(t) = e^{-at} M_0(t), \qquad (26)$$

the subscript indicating the value with respect to which the m.g.f. is constructed.

When the variate x takes only the discrete values x_i with probabilities p_i $(i = 1, 2, \ldots, n)$, the m.g.f. with respect to the value $x = a$, being the expected value of $\exp\left[t(x-a)\right]$, is given by

$$M_a(t) = \sum_i p_i \exp\left[t(x_i - a)\right] = e^{-at} M_0(t) \qquad (27)$$

as before; and the coefficient of $t^r/r!$ in the expansion is the moment of order r about the value in question. The m.g.f. will be found useful, not only in calculating the moments of a distribution, but also in leading to concise proofs of various theorems. Along with it may be mentioned the *characteristic function**** (c.f.), $E(e^{itx})$, which is the expected value of e^{itx}, where $i = \sqrt{-1}$ and t is real.

Example 1. For the *exponential distribution* defined by $\phi(x) = c e^{-cx}$, in which c is positive and x varies from 0 to ∞, the m.g.f. with respect to the origin is

$$M(t) = c \int_0^\infty \exp\left(tx - cx\right) dx \quad (|t| < c),$$

$$= (1 - t/c)^{-1} = 1 + t/c + t^2/c^2 + t^3/c^3 + \ldots,$$

showing that
$$\mu_1' = \frac{1}{c}, \quad \mu_2' = \frac{2!}{c^2}, \quad \mu_r' = \frac{r!}{c^r}.$$

Thus the mean is $1/c$, and the variance is given by

$$\mu_2 = \mu_2' - (\mu_1')^2 = \frac{1}{c^2},$$

and the s.d. by $\sigma = 1/c$.

---

* Cf. Lévy, 1925, 3, part II, chapters II and III; Cramér, 1937, 6, pp. 23–68; and Kendall, 1943, 2, pp. 90–100.

Example 2. For the *rectangular distribution*, $\phi(x) = 1/2a$, $-a \leqslant x \leqslant a$, we have

$$M_0(t) = \frac{1}{2a} \int_{-a}^{a} e^{tx} \, dx = \frac{e^{at} - e^{-at}}{2at} = \frac{\sinh at}{at}$$

$$= 1 + a^2 t^2 / 3! + a^4 t^4 / 5! + a^6 t^6 / 7! + \ldots.$$

The moments of odd order are zero, and $\mu_{2r} = a^{2r}/(2r+1)$.

Example 3. For the *binomial distribution* the m.g.f. with respect to the origin is, in virtue of (27),

$$M(t) = q^n + e^t n q^{n-1} p + e^{2t} \binom{n}{2} q^{n-2} p^2 + \cdots$$

$$= (q + p e^t)^n = (1 + pt + pt^2/2! + pt^3/3! + \ldots)^n.$$

The mean, being the coefficient of t in the expansion, is np. Similarly μ_2', being the coefficient of $t^2/2!$, has the value

$$np + \binom{n}{2} 2! p^2 = np[1 + (n-1)p].$$

Consequently the variance is given by

$$\mu_2 = \mu_2' - (\mu_1')^2 = np(1-p) = npq.$$

A very important property of the m.g.f. is expressed in the theorem:

The moment generating function of the sum of two independent variates is the product of their moment generating functions.

This is a direct consequence of the theorem concerning the expected value of the product of two independent variates. For, if x and y are independent variates, the m.g.f. of their sum with respect to the origin is

$$E(e^{t(x+y)}) = E(e^{tx} \cdot e^{ty}) = E(e^{tx}) E(e^{ty}),$$

and is therefore the product of their m.g.f.'s. And, since the origin may be chosen at pleasure, the theorem holds for the m.g.f.'s about any specified value.

Let μ_r, μ_r' denote the moments of x about the mean and the origin, and m_r, m_r' those of y. Then, by the above theorem, the m.g.f. of $x + y$ has an expansion obtained from the product

$$(1 + \mu_1' t + \mu_2' t^2/2! + \mu_3' t^3/3! + \ldots)(1 + m_1' t + m_2' t^2/2! + \ldots).$$

The coefficient of t in this product is $\mu_1' + m_1'$, so that the mean of the sum of the variates is the sum of their means. From the coefficient of $t^2/2!$ we find that the second moment of $x+y$ about the origin is $\mu_2' + 2\mu_1' m_1' + m_2'$. Consequently the second moment of $x+y$ about its mean is

$$\mu_2' + 2\mu_1' m_1' + m_2' - (\mu_1' + m_1')^2 = (\mu_2' - \mu_1'^2) + (m_2' - m_1'^2) = \mu_2 + m_2,$$

and is thus equal to the sum of the variances of x and y. Since the variance of $-y$ is equal to that of y, we have again the theorem:

The variance of the sum (or difference) of two independent variates is equal to the sum of their variances.

From their definitions it is clear that the moment generating function, when it exists, and the characteristic function, which always exists, are determined by the probability density $\phi(x)$ of the distribution. Conversely, the distribution is determined uniquely by its characteristic function,* or by the moment generating function when the moments satisfy certain conditions;† that is to say, variates which have the same moment generating function conform to the same distribution. Proofs of these statements are beyond the scope of this book. In the three cases in which we shall make use of this converse theorem (pp. 49, 57–8 and 151) the moments of the distributions satisfy the necessary conditions.

16. Cumulative function of a distribution

If the logarithm of the m.g.f. of a distribution can be expanded as a convergent series in powers of t, viz.

$$\begin{aligned} K(t) &= \log M(t) \\ &= \kappa_1 t + \kappa_2 t^2/2! + \kappa_3 t^3/3! + ..., \end{aligned} \tag{28}$$

the coefficients, κ_r, are called the *cumulants* (or *seminvariants*‡) of the distribution, and $K(t)$ is the *cumulative function*. The cumulants

---

* For a proof of this converse theorem see Lévy, 1925, 3, pp. 166–7; also Deltheil, *Erreurs et Moindres Carrés*, pp. 26–9 (Fasc. 2, Tome 1 of *Traité du Calcul des Probabilités et de ses Applications*, ed. E. Borel), Gauthier-Villars, Paris, 1930, and Kendall, 1943, 2, pp. 90–4.

† Cf. Kendall, 1943, 2, pp. 105–10.

‡ We adopt Fisher's terminology of *cumulants* (1929, 1) rather than Thiele's of *seminvariants* (1903, 1), since Dressel has shown that the cumulants are only one particular set of seminvariants (*Ann. Math. Stat.* vol. XI, pp. 33–57, 1940).

are determinate functions of the moments. For instance, on taking logarithms of both members of (24) and identifying with (28), we see that $\kappa_1 = \mu_1'$, and is thus equal to the mean of the distribution.

Since, by (26), the m.g.f. with respect to the mean is

$$M_0(t) \exp(-\mu_1' t),$$

it is clear on taking logarithms that the cumulative function with respect to the mean differs from that with respect to $x = 0$ only by the addition of the term $-\mu_1' t$. Consequently the cumulative function relative to the mean is

$$\kappa_2 t^2/2! + \kappa_3 t^3/3! + \ldots + \kappa_r t^r/r! + \ldots \tag{29}$$

And since this must be identical with

$$\log(1 + \mu_2 t^2/2! + \mu_3 t^3/3! + \ldots),$$

we see, on comparing coefficients of like powers of t, that the first few cumulants are given in terms of the moments of the distribution by the formulae

$$\kappa_1 = \mu_1' = \text{mean},$$

$$\kappa_2 = \mu_2, \quad \kappa_3 = \mu_3, \quad \kappa_4 = \mu_4 - 3\mu_2^2. \tag{30}$$

Thus all the cumulants after the first are independent of the value with respect to which the cumulative function is constructed. The mean, and the moments about the mean, may therefore be found by calculating the cumulative function with respect to any convenient origin. Also, from the last of the relations (30), we have

$$\frac{\kappa_4}{\kappa_2^2} = \frac{\mu_4}{\mu_2^2} - 3 = \beta_2 - 3, \tag{31}$$

and this is the *excess* of kurtosis, as defined at the close of § 6.

Example. Since by § 15, Ex. 1, the m.g.f. of the exponential distribution, with respect to the origin, is

$$M_0(t) = (1 - \sigma t)^{-1},$$

the cumulative function is

$$K(t) = -\log(1 - \sigma t) = \sigma t + \sigma^2 t^2/2 + \sigma^3 t^3/3 + \ldots.$$

Thus

$$\kappa_r = \sigma^r (r-1)!.$$

The mean, being the coefficient of t, is σ. The variance is σ^2, $\mu_3 = 2\sigma^3$, and

$$\mu_4 = \kappa_4 + 3\kappa_2^2 = (3! + 3)\sigma^4 = 9\sigma^4.$$

Further, since the m.g.f. of the sum of two independent variates, x and y, is equal to the product of their m.g.f.'s, it follows that the cumulative function of the distribution of $x + y$ is the sum of those of x and y. Equating coefficients of like powers of t, we have the simple result that the rth cumulant of $x + y$ is the sum of the rth cumulants of x and y. This is the *additive property of cumulants*. The theorem is obviously true for the sum of any finite number of independent variates. The particular case of second cumulants gives again the theorem on the variance of the sum of several independent variates.

The above property of cumulants may be used to estimate the average corrections* to be applied to the moments of a grouped distribution, with specified class interval c, when the interval-mesh is located at random on the ungrouped distribution. The corrections found, which are of the same form as *Sheppard's adjustments*, may be wrong in any individual instance, but their average effect in a large number of cases will be correct. In any class the mid-value x_i, from which the moments of the grouped distribution are calculated, is the sum of the true value x of the observation and the grouping error $x_i - x$. In consequence of the random location of the class limits, the grouping error is uniformly distributed over the range $-\tfrac{1}{2}c$ to $\tfrac{1}{2}c$. The average cumulants of the grouped distribution will differ from those of the ungrouped by the cumulants of the grouping error. Now the m.g.f. for the uniform distribution of error is

$$M(t) = \frac{2}{ct} \sinh\left(\tfrac{1}{2}ct\right)$$

$$= 1 + \frac{c^2}{12}\frac{t^2}{2!} + \frac{c^4}{80}\frac{t^4}{4!} + \cdots,$$

and its cumulative function is therefore

$$K(t) = \frac{c^2}{12}\frac{t^2}{2!} - \frac{c^4}{120}\frac{t^4}{4!} + \frac{c^6}{252}\frac{t^6}{6!} + \cdots.$$

If then κ'_r are the cumulants of the grouped distribution, those of the ungrouped distribution are

$$\kappa_1 = \kappa'_1, \quad \kappa_2 = \kappa'_2 - \frac{c^2}{12}, \quad \kappa_3 = \kappa'_3, \quad \kappa_4 = \kappa'_4 + \frac{c^4}{120}, \quad \text{etc.}$$

* Cf. Cornish and Fisher, 1937, 7, pp. 3–4, and Kendall, 1943, 2, pp. 74–5.

Denoting the moments of the grouped distribution by m_r and m'_r, we therefore have for those of the ungrouped distribution

$$\mu'_1 = m'_1, \quad \mu_2 = m_2 - \frac{c^2}{12}, \quad \mu_3 = m_3 \tag{32}$$

and

$$\mu_4 = \kappa_4 + 3\kappa_2^2 = \kappa'_4 + \frac{c^4}{120} + 3\left(\kappa'^2_2 - \frac{c^2\kappa'_2}{6} + \frac{c^4}{144}\right)$$

$$= m_4 - \tfrac{1}{2}c^2 m_2 + \frac{7c^4}{240}. \tag{33}$$

The equation (32) shows that the estimate m'_1 of the mean, and the estimate m_3 of the third moment about the mean, as found from the grouped distribution, are sufficiently accurate. The calculated variance m_2 should be diminished by $c^2/12$; while the adjustment to the fourth moment m_4 is given by (33).

COLLATERAL READING

USPENSKY, 1937, 2, pp. 1–16, 27–36, 44–8, 60–6, 161–72 and 235–41.
LEVY and ROTH, 1936, 1, chapters II–VI.
RIETZ, 1927, 2, chapters I and II.
AITKEN, 1939, 1, chapters I and II.
PLUMMER, 1940, 1, chapters I and II.
KENNEY, 1939, 3, part II, chapter I.
CAMP, 1934, 1, part II, chapter I.
POINCARÉ, 1912, 1, chapters I–IV, VII and IX.
COOLIDGE, 1925, 2, chapters I–IV and VI.
KENDALL, 1943, 2, chapter VII.

EXAMPLES II

1. Two cards are drawn at random from a well-shuffled pack of 52. Show that the chance of drawing two aces is $1/221$.

2. The chance of throwing a 6 at least once in two throws of a die is $11/36$.

3. A and B toss a coin alternately on the understanding that the first to obtain heads wins the toss. Show that their respective chances of winning are $2/3$ and $1/3$.

4. Four persons are chosen at random from a group containing 3 men, 2 women and 4 children. The chance that exactly two of them will be children is $10/21$.

5. From an urn containing r red and s white balls, $a+b$ balls are drawn at random without replacement $(a \leqslant r, b \leqslant s)$. Show that the probability of a red and b white balls is $\binom{r}{a}\binom{s}{b} \div \binom{r+s}{a+b}$.

6. Show that, in a single throw with two dice, the chance of throwing more than 7 is equal to that of throwing less than 7, each being $5/12$.

7. A and B take turns in throwing two dice, the first to throw 9 being awarded the prize. Show that their chances of winning are in the ratio $9:8$.

8. Three men toss in succession for a prize to be given to the one who first obtains heads. Show that their chances of winning are $4/7$, $2/7$ and $1/7$.

9. Eight coins are thrown simultaneously. Show that the chance of obtaining at least six heads is $37/256$.

10. The expectation of the number of failures preceding the first success in an indefinite series of independent trials, with constant probability p of success, is

$$qp + 2q^2p + 3q^3p + \ldots = \frac{pq}{(1-q)^2} = \frac{q}{p}.$$

(Uspensky, 1937, 2, p. 178, Ex. 3.)

11. A point P is taken at random in a line AB, of length $2a$, all positions of the point being equally likely. Show that the expected value of the area of the rectangle $AP.PB$ is $2a^2/3$, and that the probability of the area exceeding $\frac{1}{2}a^2$ is $1/\sqrt{2}$.

12. From a point on the circumference of a circle of radius a, a chord is drawn in a random direction (i.e. all directions are equally likely). Show that the expected value of the length of the chord is $4a/\pi$, and that the variance of the length is $2a^2(1 - 8/\pi^2)$. Also show that the chance is $1/3$ that the length of the chord will exceed the length of the side of an equilateral triangle inscribed in the circle.

13. A chord of a circle of radius a is drawn parallel to a given straight line, all distances from the centre of the circle being equally likely. Show that the expected value of the length of the chord is $\frac{1}{2}\pi a$, and that the variance of the length is $\frac{a^2}{12}(32-3\pi^2)$. Also show that the chance is 1/2 that the length of the chord will exceed the length of the side of an equilateral triangle inscribed in the circle.

14. Two different digits are chosen at random from the set 1, 2, 3, ..., 8. Show that the probability that the sum of the digits will be equal to 5 is the same as the probability that their sum will exceed 13, each being 1/14. Also show that the chance of both digits exceeding 5 is 3/28.

15. In Poisson's distribution the variate takes the values 0, 1, 2, 3, ... with probabilities proportional to $1, m, m^2/2!, m^3/3!, ...,$ both sequences being infinite. Show that the mean of the distribution is m, and the variance also m.

16. The equation § 15 (26), written for $a = \bar{x} = \mu_1'$, is equivalent to

$$(1+\mu_2 t^2/2! + \mu_3 t^3/3! + ...)$$
$$= (1 - \mu_1' t + \mu_1'^2 t^2/2! - ...)(1 + \mu_1' t + \mu_2' t^2/2! + ...).$$

Equating coefficients of like powers of t, deduce the relations §4 (17), (18).

17. Prove that the next two formulae corresponding to §16 (30) are

$$\kappa_5 = \mu_5 - 10\mu_3\mu_2, \quad \kappa_6 = \mu_6 - 15\mu_4\mu_2 - 10\mu_3^2 + 30\mu_2^3.$$

18. Two independent variates are each uniformly distributed within the range $-a$ to a. Show that their sum x has a probability density given by

$$\phi(x) = (2a+x)/4a^2 \quad (-2a \leqslant x \leqslant 0),$$
$$\phi(x) = (2a-x)/4a^2 \quad (0 \leqslant x \leqslant 2a).$$

Verify that the m.g.f., calculated from this value of $\phi(x)$, is equal to $\left(\frac{1}{at}\sinh at\right)^2$.

19. On the x-axis $n+1$ points are taken independently between the origin and $x = 1$, all positions being equally likely. Show that the probability that the $(k+1)$th of these points, counted from the origin, lies in the interval $x - \frac{1}{2}dx$ to $x + \frac{1}{2}dx$ is

$$\binom{n}{k}(n+1)\,x^k\,(1-x)^{n-k}\,dx.$$

Verify that the integral of this expression, from $x = 0$ to $x = 1$, is unity. (Aitken, 1939, 1, p. 71.)

20. Show that, for the binomial distribution,

$$\kappa_4 = npq(1 - 6pq).$$

21. Show that the expected value of the product of the numbers of points showing after an unbiased throw of n ordinary dice is $(7/2)^n$.

22. Show that, if p may be varied, the probability of m successes in a series of n independent trials, with the same probability p of success, is greatest when $p = m/n$.

23. Show that, if y and z are independent random values of a variate x, the expected value of $(y-z)^2$ is twice the variance of the distribution of x.

24. Defining the *harmonic mean* (H.M.) of a variate x as the reciprocal of the expected value of $1/x$, show that the H.M. of the variate which ranges from 0 to ∞ with probability density $x^n e^{-x}/n!$ is n, given that n is positive.

CHAPTER III

SOME STANDARD DISTRIBUTIONS

BINOMIAL AND POISSONIAN DISTRIBUTIONS

17. The binomial distribution

We have considered the binomial distribution in connection with the probabilities of the various numbers of successes in a series of n independent trials, in each of which the chance of success is equal to p. The mean and the variance of the distribution were determined by means of the property that the expected value of a sum of variates is equal to the sum of their expected values. These may also be found by direct calculation. Thus

$$E(x) = 0 \cdot q^n + 1 \cdot nq^{n-1}p + 2 \cdot \binom{n}{2} q^{n-2}p^2 + \ldots + np^n$$

$$= np\left[q^{n-1} + (n-1)q^{n-2}p + \binom{n-1}{2} q^{n-3}p^2 + \ldots + p^{n-1} \right]$$

$$= np(q+p)^{n-1} = np. \tag{1}$$

Thus the mean of the distribution is np. In order to find the variance calculate first the second moment about $x = 0$. Thus

$$\mu_2' = 0^2 \cdot q^n + 1^2 \cdot nq^{n-1}p + 2^2 \cdot \binom{n}{2} q^{n-2}p^2 + \ldots + n^2 \cdot p^n$$

$$= np\left[q^{n-1} + 2(n-1)q^{n-2}p + 3\binom{n-1}{2} q^{n-3}p^2 + \ldots + np^{n-1} \right]$$

Now the expression in brackets is the first moment, about the value $x = -1$, for the binomial distribution in which n is replaced by $n-1$. This first moment, being the excess of the mean of the distribution above $x = -1$, is equal to $(n-1)p + 1$, in virtue of (1). Consequently

$$\mu_2' = np[(n-1)p + 1],$$

and the variance of the binomial distribution is then given by

$$\mu_2 = \mu_2' - [E(x)]^2 = np[1 + (n-1)p] - (np)^2$$

$$= np(1-p) = npq, \tag{2}$$

and the standard deviation by

$$\sigma = \sqrt{(npq)}. \tag{3}$$

The proportion or *relative frequency of successes* is the number of successes divided by n. Hence the mean value of this proportion is p, and its standard deviation is $\sqrt{(pq/n)}$.

A binomial frequency distribution is one in which the relative frequencies of the values $0, 1, 2, \ldots, n$ of the variate are equal to their probabilities in the above distribution. As an example we may take the distribution of expected frequencies of $0, 1, \ldots, n$ successes when the set of n trials is to be made N times. For, in virtue of (1) and the known probability of r successes, the expected frequency of r successes in N sets of n trials each is $N\binom{n}{r}p^r q^{n-r}$. The properties of the distribution are the same whether it is regarded as one of probability or one of frequency. But the reader may note that, in a theoretical frequency distribution like the above, the individual frequencies are not necessarily integral.

Example 1. Verify the above value of μ_2' by direct algebraical simplification of the expression.

Example 2. Show that the m.g.f. of the binomial distribution with respect to its mean is

$$(qe^{-pt}+pe^{qt})^n = \left[1+pq\frac{t^2}{2!}+pq(q^2-p^2)\frac{t^3}{3!}+pq(q^3+p^3)\frac{t^4}{4!}+\ldots \right]^n,$$

and deduce that

$$\mu_2 = npq, \quad \mu_3 = npq(q-p), \quad \mu_4 = npq[1+3(n-2)pq].$$

18. Poisson's distribution

An important distribution, associated with the name of Poisson, is one obtainable from the binomial distribution by putting $p = m/n$, where m is a constant, and letting n increase indefinitely. Thus the number of trials in the series becomes very large, and the probability of success in a trial very small. Now it can be shown* that, on the above assumption as n tends to infinity,

$$\lim \binom{n}{r}p^r q^{n-r} = \frac{m^r e^{-m}}{r!}.$$

* See Mathematical Note I, at the end of this chapter.

Thus, in the limiting form of the distribution, the probability of r successes in the infinite series of trials is $m^r e^{-m}/r!$. The chances of $0, 1, 2, 3, \ldots$ successes in the infinite series are

$$e^{-m}, \quad e^{-m}m, \quad e^{-m}m^2/2!, \quad e^{-m}m^3/3!, \quad \ldots \tag{4}$$

respectively. The reader may show, as in the case of the binomial distribution, that the most probable number of successes is the integral part of m. It is obvious that the sum of the probabilities of $0, 1, 2, \ldots$ successes is $e^{-m}e^m = 1$, as it should be.

The mean value and the variance of Poisson's distribution may be deduced from those of the binomial by putting $p = m/n$, and letting n tend to infinity. Thus the *mean value* is $\lim np = m$. Similarly, the *variance* is $\lim npq = \lim mq = m$, since q tends to unity as p tends to zero. These results may, of course, be deduced by direct calculation. Thus the mean, being the first moment about $x = 0$, is given by

$$\begin{aligned}
\bar{x} = \mu_1' &= e^{-m}(0 + 1 \cdot m + 2 \cdot m^2/2! + 3 \cdot m^3/3! + \ldots) \\
&= me^{-m}(1 + m + m^2/2! + m^3/3! + \ldots) \\
&= m.
\end{aligned} \tag{5}$$

Similarly
$$\begin{aligned}
\mu_2' &= e^{-m}(0^2 + 1^2 \cdot m + 2^2 \cdot m^2/2! + 3^2 \cdot m^3/3! + \ldots) \\
&= m(m+1),
\end{aligned}$$

as the reader may easily verify. Consequently

$$\begin{aligned}
\sigma^2 = \mu_2 = \mu_2' - \bar{x}^2 &= m(m+1) - m^2 \\
&= m,
\end{aligned} \tag{6}$$

as found above. The s.d. is therefore $\sqrt{m}$.

The same results may be obtained by using the generating functions of §§ 15 and 16. Thus the m.g.f. of the Poissonian distribution with respect to the origin is

$$\begin{aligned}
M_0(t) &= e^{-m}(e^{0t} + e^t m + e^{2t}m^2/2! + e^{3t}m^3/3! + \ldots) \\
&= e^{-m}\exp(me^t) = \exp[m(e^t - 1)],
\end{aligned}$$

and the cumulative function is therefore

$$K(t) = m(e^t - 1) = m(t + t^2/2! + t^3/3! + \ldots).$$

Thus $\kappa_1 = \kappa_2 = \kappa_3 = \ldots = m$. The mean and the variance are each equal to m, as is also the third moment about the mean. The fourth moment about the mean is

$$\mu_4 = \kappa_4 + 3\kappa_2^2 = m + 3m^2.$$

Further, if n independent variates x_i conform to Poissonian distributions with means m_i $(i = 1, 2, \ldots, n)$, it follows from the theorem of §15 that the m.g.f. of their sum is $\exp\left[(e^t - 1) \sum_i m_i\right]$. But this is the m.g.f. of a Poissonian distribution whose mean is $\sum_i m_i$. Hence the theorem:

*The sum of any finite number of independent Poissonian variates is itself a Poissonian variate, with mean equal to the sum of the means of the separate variates.**

A Poissonian frequency distribution is one in which the relative frequencies of the values 0, 1, 2, 3, ... of the variate are equal to the probabilities in the above distribution. As an example we have the distribution of the expected frequencies of the various numbers of successes when the extensive series of trials is repeated N times. But in such a theoretical distribution the individual frequencies are not necessarily integral. Frequency distributions which are approximately Poissonian do arise in connection with the number of happenings of a rare event in an extensive series of trials.

Example. In 1,000 consecutive issues of *The Utopian Seven-daily Chronicle* the deaths of centenarians were recorded,† the number x having frequency f according to the table

x:	0	1	2	3	4	5	6	7	8
f:	229	325	257	119	50	17	2	1	0

Show that the distribution is roughly Poissonian by calculating its mean ($m = 1\cdot5$), and then the frequencies in the Poissonian distribution with the same mean and the same total frequency of 1,000. The latter are approximately 223·1, 334·7, 251·0, 125·5, 47·1, 14·1, 3·5, 0·8, 0·2. Also calculate the variance of the given distribution, and compare it with the mean.

---

* An elementary proof of this theorem will be found in Ex. 4 at the end of this chapter.

† Cf. Lucy Whittaker, *Biometrika*, vol. x, 1914, p. 36.

THE NORMAL DISTRIBUTION

19. Derivation from the binomial distribution

The binomial and Poissonian are distributions of discrete values. We pass now to a continuous distribution of fundamental importance. This is the *normal distribution*, which may be derived from the binomial in the following manner. In the latter the probability of the value r of the variate is $\binom{n}{r} p^r q^{n-r}$. Let x be the deviation of the variate from the mean value np of r, so that

$$r = np + x. \tag{7}$$

Then the probability of the value x, being the same as that of the corresponding value of r, is

$$P = \binom{n}{r} p^r q^{n-r} = \binom{n}{np+x} p^{np+x} q^{nq-x}.$$

It can be shown* that, for large values of n, this probability can be expressed

$$P = \frac{1}{\sqrt{(2\pi npq)}} \exp\left(-\frac{x^2}{2npq} + \epsilon\right), \tag{8}$$

where ϵ tends to zero as n tends to infinity, provided that neither p nor q is very small, and x is of lower order than $n^{2/3}$.

Now introduce a variable z defined by

$$z = \frac{x}{\sqrt{n}},$$

and examine the probability distribution of z, and its limiting form as n tends to infinity. The probability for any interval in the range of z is equal to that of the corresponding interval for x. To unit interval in the range of x there corresponds the interval $1/\sqrt{n}$ in that of z; and, when n tends to infinity, this may be denoted by dz. Then the probability that z falls in the interval dz is the probability that x falls in unit interval which includes the value x; and this is given by (8) which, when n tends to infinity, takes the form

$$dP = \frac{dz}{\sqrt{(2\pi pq)}} \exp\left(-\frac{z^2}{2pq}\right).$$

* For the details of this step see Mathematical Note II, at the end of this chapter.

The limiting form of the distribution of z is thus a continuous distribution with probability density

$$\phi(z) = \frac{1}{\sigma\sqrt{(2\pi)}}\exp\left(-\frac{z^2}{2\sigma^2}\right), \tag{9}$$

where $\sigma^2 = pq$. This is the normal probability function, and the corresponding continuous distribution is the normal distribution. Since $\sqrt{(npq)}$ is the S.D. of the binomial distribution, and therefore of the variate x, it follows that $\sqrt{(pq)}$ is the S.D. of z. Hence σ in (9) is the S.D. of the normal distribution. Since $\phi(z)$ is an even function of z, the distribution is *symmetrical* about $z = 0$, which is therefore the *mean* value of the distribution. Also, since z must lie between $-\infty$ and ∞, the integral of (9) between these limits is equal to unity, so that

$$\int_{-\infty}^{\infty}\exp\left(-\frac{z^2}{2\sigma^2}\right)dz = \sigma\sqrt{(2\pi)}. \tag{10}$$

This is an important integral. An independent proof of the formula is given in Mathematical Note III, at the end of this chapter.

A normal frequency distribution is a continuous one, in which the relative frequency density $f(x)$ is identical with the function $\phi(x)$ defined by (9). A distribution of discrete values cannot be normal. If, however, the total frequency is large, it is possible for the discrete distribution to approximate to the normal. The meaning of such an approximation was explained in § 6.

In later chapters, particularly in connection with sampling theory, we shall meet distributions which are accurately normal, and others which are approximately so. Since the normal distribution is an ideal to which some distributions attain, and to which others approximate, it is important to know its properties. Fortunately, the quantities associated with the distribution are easy to calculate.

20. Some properties of the normal distribution

As we have just seen, a continuous variate x is normally distributed, with mean zero and S.D. σ, when the range of the variate is from $-\infty$ to ∞, and the probability density is given by

$$\phi(x) = \frac{1}{\sigma\sqrt{(2\pi)}}\exp\left(-\frac{x^2}{2\sigma^2}\right). \tag{11}$$

The probability curve is therefore the curve

$$y = \frac{1}{\sigma\sqrt{(2\pi)}}\exp\left(-\frac{x^2}{2\sigma^2}\right).$$ (12)

This curve is *symmetrical* about the line $x = 0$ through the mean of the distribution. It is a uni-modal curve, in which the ordinates decrease rapidly as $|x|$ increases. By equating to zero the second derivative of y, the reader will easily verify that the points of inflexion on the curve are given by $x = \pm\sigma$. The ordinates of the normal curve are given in Table 1, corresponding to values of x/σ at intervals of 0·01.

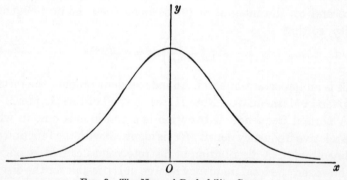

FIG. 3. The Normal Probability Curve

Since the distribution is symmetrical, the moments of odd order about the mean are all zero. To find the *moments of even order* we observe that the moment of order $2n$ about the mean is

$$\mu_{2n} = \int_{-\infty}^{\infty} x^{2n}\phi(x)\,dx = \frac{1}{\sigma\sqrt{(2\pi)}}\int_{-\infty}^{\infty} x^{2n-1}x\exp\left(-\frac{x^2}{2\sigma^2}\right)dx.$$

Integration by parts then gives

$$\mu_{2n} = -\frac{\sigma}{\sqrt{(2\pi)}}\left[x^{2n-1}\exp\left(-\frac{x^2}{2\sigma^2}\right)\right]_{-\infty}^{\infty} + \frac{\sigma(2n-1)}{\sqrt{(2\pi)}}\int_{-\infty}^{\infty} x^{2n-2}\exp\left(-\frac{x^2}{2\sigma^2}\right)dx.$$

The expression in square brackets vanishes at both limits; and we may therefore write the relation

$$\mu_{2n} = (2n-1)\sigma^2\mu_{2n-2}.$$

TABLE 1. *Ordinates of the Normal Curve*

The origin of x is at the mean. The table gives the values of $\dfrac{1}{\sqrt{(2\pi)}} \exp\left(-\dfrac{x^2}{2\sigma^2}\right)$,

which is σ times the ordinate of the normal curve $y = \phi(x)$

$\dfrac{x}{\sigma}$	0·00	0·01	0·02	0·03	0·04	0·05	0·06	0·07	0·08	0·09
0·0	·3989	·3989	·3989	·3988	·3986	·3984	·3982	·3980	·3977	·3973
0·1	·3970	·3965	·3961	·3956	·3951	·3945	·3939	·3932	·3925	·3918
0·2	·3910	·3902	·3894	·3885	·3876	·3867	·3857	·3847	·3836	·3825
0·3	·3814	·3802	·3790	·3778	·3765	·3752	·3739	·3725	·3712	·3697
0·4	·3683	·3668	·3653	·3637	·3621	·3605	·3589	·3572	·3555	·3538
0·5	·3521	·3503	·3485	·3467	·3448	·3429	·3410	·3391	·3372	·3352
0·6	·3332	·3312	·3292	·3271	·3251	·3230	·3209	·3187	·3166	·3144
0·7	·3123	·3101	·3079	·3056	·3034	·3011	·2989	·2966	·2943	·2920
0·8	·2897	·2874	·2850	·2827	·2803	·2780	·2756	·2732	·2709	·2685
0·9	·2661	·2637	·2613	·2589	·2565	·2541	·2516	·2492	·2468	·2444
1·0	·2420	·2396	·2371	·2347	·2323	·2299	·2275	·2251	·2227	·2203
1·1	·2179	·2155	·2131	·2107	·2083	·2059	·2036	·2012	·1989	·1965
1·2	·1942	·1919	·1895	·1872	·1849	·1826	·1804	·1781	·1758	·1736
1·3	·1714	·1691	·1669	·1647	·1626	·1604	·1582	·1561	·1539	·1518
1·4	·1497	·1476	·1456	·1435	·1415	·1394	·1374	·1354	·1334	·1315
1·5	·1295	·1276	·1257	·1238	·1219	·1200	·1182	·1163	·1145	·1127
1·6	·1109	·1092	·1074	·1057	·1040	·1023	·1006	·0989	·0973	·0957
1·7	·0940	·0925	·0909	·0893	·0878	·0863	·0848	·0833	·0818	·0804
1·8	·0790	·0775	·0761	·0748	·0734	·0721	·0707	·0694	·0681	·0669
1·9	·0656	·0644	·0632	·0620	·0608	·0596	·0584	·0573	·0562	·0551
2·0	·0540	·0529	·0519	·0508	·0498	·0488	·0478	·0468	·0459	·0449
2·1	·0440	·0431	·0422	·0413	·0404	·0395	·0387	·0379	·0371	·0363
2·2	·0355	·0347	·0339	·0332	·0325	·0317	·0310	·0303	·0297	·0290
2·3	·0283	·0277	·0270	·0264	·0258	·0252	·0246	·0241	·0235	·0229
2·4	·0224	·0219	·0213	·0208	·0203	·0198	·0194	·0189	·0184	·0180
2·5	·0175	·0171	·0167	·0163	·0158	·0154	·0151	·0147	·0143	·0139
2·6	·0136	·0132	·0129	·0126	·0122	·0119	·0116	·0113	·0110	·0107
2·7	·0104	·0101	·0099	·0096	·0093	·0091	·0088	·0086	·0084	·0081
2·8	·0079	·0077	·0075	·0073	·0071	·0069	·0067	·0065	·0063	·0061
2·9	·0060	·0058	·0056	·0055	·0053	·0051	·0050	·0048	·0047	·0046
3·0	·0044	·0043	·0042	·0040	·0039	·0038	·0037	·0036	·0035	·0034
3·1	·0033	·0032	·0031	·0030	·0029	·0028	·0027	·0026	·0025	·0025
3·2	·0024	·0023	·0022	·0022	·0021	·0020	·0020	·0019	·0018	·0018
3·3	·0017	·0017	·0016	·0016	·0015	·0015	·0014	·0014	·0013	·0013
3·4	·0012	·0012	·0012	·0011	·0011	·0010	·0010	·0010	·0009	·0009
3·5	·0009	·0008	·0008	·0008	·0008	·0007	·0007	·0007	·0007	·0006
3·6	·0006	·0006	·0006	·0005	·0005	·0005	·0005	·0005	·0005	·0004
3·7	·0004	·0004	·0004	·0004	·0004	·0004	·0003	·0003	·0003	·0003
3·8	·0003	·0003	·0003	·0003	·0003	·0002	·0002	·0002	·0002	·0002
3·9	·0002	·0002	·0002	·0002	·0002	·0002	·0002	·0002	·0001	·0001

Repeated application of this reduction formula shows that

$$\mu_{2n} = (2n-1)(2n-3)\ldots 3 . 1 . \sigma^{2n}\mu_0.$$

But, for any distribution, $\mu_0 = 1$, and therefore

$$\mu_{2n} = (2n-1)(2n-3)\ldots 3 . 1 . \sigma^{2n}. \tag{13}$$

In particular, the variance is σ^2 and the s.d. is σ, as proved above. Similarly,

$$\mu_4 = 3\sigma^4. \tag{14}$$

The moments may also be obtained from the moment generating function of the normal distribution. This function, relative to the mean $x = 0$, is

$$\begin{aligned} M(t) &= \int_{-\infty}^{\infty} e^{tx}\phi(x)\,dx = \frac{1}{\sigma\sqrt{(2\pi)}}\int_{-\infty}^{\infty} \exp\left(tx - \frac{x^2}{2\sigma^2}\right) dx \\ &= \frac{1}{\sigma\sqrt{(2\pi)}}\exp\left(\tfrac{1}{2}\sigma^2 t^2\right)\int_{-\infty}^{\infty} \exp\left[-\frac{1}{2\sigma^2}(x - \sigma^2 t)^2\right] dx \\ &= \exp\left(\tfrac{1}{2}\sigma^2 t^2\right), \end{aligned} \tag{15}$$

in virtue of (10). Since the expansion of this expression involves only even powers of t, the moments of odd order about the mean are zero, as is obvious from symmetry. The moment of order $2n$ is the coefficient of $t^{2n}/(2n)!$ in the expansion; and this has the value

$$\mu_{2n} = (\tfrac{1}{2}\sigma^2)^n (2n)!/n! = 1 . 3 . 5 \ldots (2n-1)\sigma^{2n}$$

as found above. The cumulative function is, in virtue of (15),

$$K(t) = \log M(t) = \tfrac{1}{2}\sigma^2 t^2,$$

so that all the cumulants after the second are equal to zero.

The *mean deviation* from the mean is easily calculated. Its value is

$$\int_{-\infty}^{\infty} |x|\,\phi(x)\,dx = 2\int_{0}^{\infty} x\phi(x)\,dx$$

in virtue of symmetry; and this integral

$$\begin{aligned} &= \frac{\sqrt{2}}{\sigma\sqrt{\pi}}\int_{0}^{\infty} x\exp\left(-\frac{x^2}{2\sigma^2}\right) dx = -\frac{\sigma\sqrt{2}}{\sqrt{\pi}}\left[\exp\left(-\frac{x^2}{2\sigma^2}\right)\right]_{0}^{\infty} \\ &= \sigma\sqrt{\frac{2}{\pi}} = 0{\cdot}7979\sigma = \tfrac{4}{5}\sigma \text{ approximately.} \end{aligned}$$

In a normal distribution with mean $\bar{x}$ and s.d. σ, $x - \bar{x}$ is the deviation of the variable from the mean, and the probability density is

$$\phi(x) = \frac{1}{\sigma \sqrt{(2\pi)}} \exp\left[-\frac{(x - \bar{x})^2}{2\sigma^2}\right]. \tag{16}$$

Conversely, a probability density of this form defines a normal distribution with mean $\bar{x}$ and s.d. σ.

21. Probabilities and relative frequencies for various intervals

The probability corresponding to any interval in the range of the variate is represented by the area under the curve (12) within that interval; and the same is true for the relative frequency in the case of a normal frequency distribution. In particular, the probability for the interval from the mean (zero) to the value x is given by the integral

$$P = \frac{1}{\sigma \sqrt{(2\pi)}} \int_0^x \exp\left(-\frac{x^2}{2\sigma^2}\right) dx.$$

Putting $t = x/\sigma$ we see that this is equivalent to

$$P = \frac{1}{\sqrt{(2\pi)}} \int_0^t \exp\left(-\tfrac{1}{2}t^2\right) dt. \tag{17}$$

This probability is therefore a function of x/σ. In the accompanying table the values of this integral are given for different values of x/σ at intervals of 0·01.

Using Table 2 the reader will see that, if $x/\sigma = 1$, the area is 0·3413. The area within the interval $x = -\sigma$ to $x = \sigma$ is therefore 0·6826, and the area outside this interval 0·3174, which is less than 1/3. In other words, the probability that a random value of a normal variate will deviate more than σ from the mean, is less than 1/3. Similarly, for $x/\sigma = 2$ the value of the integral (17) is 0·4772, and the area for the interval $x = -2\sigma$ to $x = 2\sigma$ is 0·9544. The area outside this interval is thus 0·0456. The probability that a random value of x will deviate more than 2σ from the mean is thus about $4\tfrac{1}{2}$ %. Similarly, the area outside the range $x = -2\cdot5\sigma$ to $x = 2\cdot5\sigma$ is 0·0124, which is about $1\tfrac{1}{4}$ % of the whole; and that outside the range $x = -3\sigma$ to $x = 3\sigma$ is only 0·0027, or about $\tfrac{1}{4}$ % of the whole. The reader may verify, and will find it convenient to remember, that the deviation from the mean which is exceeded with a probability of 5 % is 1·96σ; and that which is exceeded with a probability of 1 % is 2·58σ.

TABLE 2. Area under the Normal Curve

The area is measured from the mean, $x = 0$, to any ordinate, $x = \varpi$.
The results are given for values of x/σ at intervals of 0·01

$\dfrac{x}{\sigma}$	0·00	0·01	0·02	0·03	0·04	0·05	0·06	0·07	0·08	0·09
0·0	·0000	·0040	·0080	·0120	·0159	·0199	·0239	·0279	·0319	·0359
0·1	·0398	·0438	·0478	·0517	·0557	·0596	·0636	·0675	·0714	·0753
0·2	·0793	·0832	·0871	·0910	·0948	·0987	·1026	·1064	·1103	·1141
0·3	·1179	·1217	·1255	·1293	·1331	·1368	·1406	·1443	·1480	·1517
0·4	·1554	·1591	·1628	·1664	·1700	·1736	·1772	·1808	·1844	·1879
0·5	·1915	·1950	·1985	·2019	·2054	·2088	·2123	·2157	·2190	·2224
0·6	·2257	·2291	·2324	·2357	·2389	·2422	·2454	·2486	·2518	·2549
0·7	·2580	·2611	·2642	·2673	·2704	·2734	·2764	·2794	·2823	·2852
0·8	·2881	·2910	·2939	·2967	·2995	·3023	·3051	·3078	·3106	·3133
0·9	·3159	·3186	·3212	·3238	·3264	·3289	·3315	·3340	·3365	·3389
1·0	·3413	·3438	·3461	·3485	·3508	·3531	·3554	·3577	·3599	·3621
1·1	·3643	·3665	·3686	·3708	·3729	·3749	·3770	·3790	·3810	·3830
1·2	·3849	·3869	·3888	·3907	·3925	·3944	·3962	·3980	·3997	·4015
1·3	·4032	·4049	·4066	·4082	·4099	·4115	·4131	·4147	·4162	·4177
1·4	·4192	·4207	·4222	·4236	·4251	·4265	·4279	·4292	·4306	·4319
1·5	·4332	·4345	·4357	·4370	·4382	·4394	·4406	·4418	·4430	·4441
1·6	·4452	·4463	·4474	·4485	·4495	·4505	·4515	·4525	·4535	·4545
1·7	·4554	·4564	·4573	·4582	·4591	·4599	·4608	·4616	·4625	·4633
1·8	·4641	·4649	·4656	·4664	·4671	·4678	·4686	·4693	·4699	·4706
1·9	·4713	·4719	·4726	·4732	·4738	·4744	·4750	·4756	·4762	·4767
2·0	·4772	·4778	·4783	·4788	·4793	·4798	·4803	·4808	·4812	·4817
2·1	·4821	·4826	·4830	·4834	·4838	·4842	·4846	·4850	·4854	·4857
2·2	·4861	·4865	·4868	·4871	·4875	·4878	·4881	·4884	·4887	·4890
2·3	·4893	·4896	·4898	·4901	·4904	·4906	·4909	·4911	·4913	·4916
2·4	·4918	·4920	·4922	·4925	·4927	·4929	·4931	·4932	·4934	·4936
2·5	·4938	·4940	·4941	·4943	·4945	·4946	·4948	·4949	·4951	·4952
2·6	·4953	·4955	·4956	·4957	·4959	·4960	·4961	·4962	·4963	·4964
2·7	·4965	·4966	·4967	·4968	·4969	·4970	·4971	·4972	·4973	·4974
2·8	·4974	·4975	·4976	·4977	·4977	·4978	·4979	·4980	·4980	·4981
2·9	·4981	·4982	·4983	·4983	·4984	·4984	·4985	·4985	·4986	·4986
3·0	·49865	·4987	·4987	·4988	·4988	·4989	·4989	·4989	·4990	·4990
3·1	·49903	·4991	·4991	·4991	·4992	·4992	·4992	·4992	·4993	·4993

From Table 2 we may find the *quartiles* of the normal distribution with mean at $x = 0$. The relative frequency from the mean to the upper quartile is 0·25, which is therefore the corresponding area under the curve. Interpolating with the aid of Table 2 we find

$x/\sigma = 0.6745$, and the upper quartile is therefore 0.6745σ. Similarly, the lower quartile is -0.6745σ.

Example 1. Using Table 2 show that the 6th, 7th, 8th and 9th deciles are 0.253σ, 0.525σ, 0.842σ and 1.282σ respectively.

Example 2. For a normal distribution, with mean $\bar{x} = 1$ and s.d. 3, find the probabilities for the intervals

(i) $x = 3.43$ to $x = 6.19$, (ii) $x = -1.43$ to $x = 6.19$.

Let x' denote the deviation from the mean. Then, in the first part of the example, the values of x'/σ for the bounds of the interval are 0.81 and 1.73. The areas from the mean to the bounds of the interval are 0.2910 and 0.4582. The area corresponding to the interval is the difference of these areas, and is therefore 0.1672.

In the second part the values of x'/σ corresponding to the bounds of the interval are -0.81 and 1.73. The area from the lower bound to the mean is 0.2910, and that from the mean to the upper bound is 0.4582 as before. The required area is the sum of these, viz. 0.7492.

22. Distribution of a sum of independent normal variates

Consider the normal variate x, with mean a and variance σ^2. Its m.g.f. with respect to the origin is

$$M(t) = E(e^{tx}) = \frac{1}{\sigma\sqrt{(2\pi)}} \int_{-\infty}^{\infty} \exp\left[tx - \frac{(x-a)^2}{2\sigma^2}\right] dx$$

$$= \frac{1}{\sigma\sqrt{(2\pi)}} \exp\left(at + \tfrac{1}{2}\sigma^2 t^2\right) \int_{-\infty}^{\infty} \exp\left[-\frac{1}{2\sigma^2}(x - a - \sigma^2 t)^2\right] dx$$

$$= \exp\left(at + \tfrac{1}{2}\sigma^2 t^2\right), \tag{18}$$

in virtue of (10). The cumulative function is

$$K(t) = \log M(t) = at + \tfrac{1}{2}\sigma^2 t^2.$$

If c is a constant, cx is normally distributed with mean ca and variance $c^2\sigma^2$. Hence the m.g.f. of the distribution of cx is

$$\exp\left(cat + \tfrac{1}{2}c^2\sigma^2 t^2\right).$$

Let x_i $(i = 1, ..., n)$ be n independent normal variates, with means a_i and variances σ_i^2. Then, by the theorem of § 15, the m.g.f. of the sum $\sum c_i x_i$ is

$$\exp\left[t \sum c_i a_i + \tfrac{1}{2}t^2 \sum c_i^2 \sigma_i^2\right].$$

But this is the m.g.f. of a normal distribution with mean $\sum c_i a_i$ and variance $\sum c_i^2 \sigma_i^2$. Hence the theorem:

If the independent variates x_i $(i = 1, ..., n)$ are normally distributed with means a_i and variances σ_i^2, the variate $\sum c_i x_i$ is normally distributed with mean $\sum c_i a_i$ and variance $\sum c_i^2 \sigma_i^2$.

In particular, the sum (or the difference) of two independent normal variates is normally distributed, with variance equal to the sum of their variances. Also, if in the above theorem we put

$$a_i = a, \quad \sigma_i = \sigma, \quad c_i = 1/n \quad (i = 1, 2, ..., n),$$

we have the important result:

If the independent variates x_i $(i = 1, ..., n)$ are normally distributed about a common mean, a, with a common variance, σ^2, their mean is also normally distributed about a, but with variance σ^2/n.

Other proofs of the above theorem will be found in Ex. 8 at the end of this chapter.

COLLATERAL READING

YULE and KENDALL, 1937, 1, chapter x.
KENNEY, 1939, 3, part I, chapter VI.
AITKEN, 1939, 1, chapter III.
JONES, 1924, 3, chapter XVIII.
CAMP, 1934, 1, part I, chapters V and VI.
FISHER, 1938, 2, chapter III.
RIDER, 1939, 6, chapter V (§§ 30–32)
PLUMMER, 1940, 1, chapter III.
MILLS, 1938, 1, chapter XIII.
GOULDEN, 1939, 2, chapter III.
RIETZ (ed.), 1924, 1, chapter VII (to p. 103).
BOWLEY, 1920, 2, part II, chapters II and III.
COOLIDGE, 1925, 2, chapter VII.
LEVY and ROTH, 1936, 1, chapter VIII.

EXAMPLES III

1. Show by direct calculation that the third moment of the binomial distribution about $x = 0$ is

$$\mu_3' = np[(n-1)(n-2)p^2 + 3(n-1)p + 1],$$

and deduce that $\mu_3 = npq(q-p)$. Similarly, show that

$$\mu_4 = npq[1 + 3(n-2)pq].$$

2. Show that, for Poisson's distribution, the third and fourth moments about $x = 0$ are given by

$$\mu_3' = m[(m+1)^2 + m], \quad \mu_4' = m(m^3 + 6m^2 + 7m + 1),$$

and deduce that

$$\mu_3 = m, \quad \mu_4 = 3m^2 + m.$$

Deduce these results also as limiting values of the corresponding moments in Ex. 1.

3. In a normal distribution, whose mean is 2 and S.D. 3, find a value of the variate such that the probability of the interval from the mean to that value is 0·4115. (*Ans. $x = 6 \cdot 05$.*) Find another value such that the probability for the interval from $x = 3 \cdot 5$ to that value is 0·2307. (*Ans. $x = 6 \cdot 26$.*)

4. *If two independent variates, x and y, have Poissonian distributions with means m_1 and m_2, their sum is a Poissonian variate with mean $m_1 + m_2$.* (Cf. § 18.)

The variates take only the values 0, 1, 2, 3, ... We require the probability that their sum will take the value r. The probability that simultaneously x will have the value s and y the value $r - s$ is

$$\frac{m_1^s \exp(-m_1)}{s!} \times \frac{m_2^{r-s} \exp(-m_2)}{(r-s)!}.$$

Summing for all values of s from 0 to r, we see that the probability that $x + y$ will have the value r is

$$\exp(-m_1 - m_2) \sum_{s=0}^{r} \frac{m_1^s m_2^{r-s}}{s!(r-s)!} = \frac{(m_1 + m_2)^r \exp(-m_1 - m_2)}{r!}.$$

Consequently $x + y$ is a Poissonian variate with mean $m_1 + m_2$. It follows that, if the independent variates x_i have Poissonian distributions with means m_i ($i = 1, ..., n$), their sum is a Poissonian variate with mean $\sum m_i$.

5. Consider the normal distribution which has the same mean (3·95 lb.) and the same S.D. (0·46 lb.) as the distribution of yields of grain in Ex. I, 2. Find the relative frequencies of the normal distribution for the intervals corresponding to the various classes, and deduce the class frequencies per 500 of the normal distribution. This process is sometimes spoken of as *fitting a normal distribution* to the data.

Take, for example, the class whose limits are 4·1 and 4·3 lb. Since the mean is 3·95 we have, for the lower limit, $x/\sigma = 0.15/0.46 = 0.3261$. The area from the mean to this limit is therefore 0·1278, and the area to the right of it is 0·3722. Similarly, for the upper limit $x/\sigma = 0.35/0.46 = 0.7609$; and the area to the right of this is 0·2233. The difference of the areas to the right of the limits is that corresponding to the interval. Its value is 0·1489. This is the relative frequency for that interval.

The work can be tabulated as below. Here x denotes the deviation of a class limit from the mean 3·95 lb. Knowing that $\sigma = 0.46$ lb. we can find x/σ for each limit. The third column gives the area under the normal curve, to the right of the class limit. The differences, recorded in the next column, are the relative frequencies for the various classes. These differences, multiplied by 500, give the normal frequencies of the classes per 500 of the total. The last column records the class frequencies in the given distribution. The lower limit of the lowest class is taken as $-\infty$, and the upper limit of the highest class as ∞, so as to include the whole of the normal distribution.

Fitting a normal distribution to that of 500 yields of grain

Class limit	x/σ	Area to right	Difference d	$500d$	Observed f
$-\infty$	$-\infty$	1·0000			
2·9	-2.2826	0·9888	0·0112	5·6	4
3·1	-1.8479	0·9676	0·0212	10·6	15
3·3	-1.4130	0·9212	0·0464	23·2	20
3·5	-0.9783	0·8361	0·0851	42·6	47
3·7	-0.5435	0·7066	0·1295	64·7	63
3·9	-0.1087	0·5433	0·1633	81·6	78
4·1	0·3261	0·3722	0·1712	85·6	88
4·3	0·7609	0·2233	0·1489	74·4	69
4·5	1·1956	0·1160	0·1073	53·7	59
4·7	1·6304	0·0516	0·0644	32·2	35
4·9	2·0652	0·0194	0·0322	16·1	10
5·1	2·5000	0·0062	0·0132	6·6	8
∞	∞	0·0000	0·0062	3·1	4

6. As in the previous example, fit a normal distribution to the distribution of wages of 1,000 employees given in Ex. I, 3, showing that the class frequencies per thousand of the normal distribution are approximately 6·7, 11·3, 25·0, 48·0, 79·0, 113·1, 140·5, 151·0, 140·8, 113·5, 79·5, 48·1, 25·3, 11·5 and 6·7.

7. In 1,000 extensive sets of trials for an event of small probability, the frequencies f_i of the numbers x_i of successes proved to be

x:	0	1	2	3	4	5	6	7
f:	305	365	210	80	28	9	2	1

Show that the mean number of successes is 1·2, and hence that the frequencies of the Poissonian distribution, with the same mean and the same total frequency, are approximately 301·2, 361·4, 216·8, 86·7, 26·0, 6·2, 1·2, 0·2.... Verify that the variance of the given distribution is 1·28.

8. *Sum of independent normal variates.* Another proof of the theorem of § 22 may be given as follows. Let x and y be the independent normal variates, with a common mean zero and variances σ_1^2 and σ_2^2 respectively; and let $u = x + y$. Then, for a fixed value of y, $du = dx$. The probability that the value of x will fall in the interval dx is $\phi(x)\,dx$; and therefore, for a fixed value of y in the interval dy, the probability that u will fall in the interval du is

$$dp_1 = \frac{du}{\sigma_1 \sqrt{(2\pi)}} \exp\left[-\frac{(u-y)^2}{2\sigma_1^2} \right].$$

But the probability of y falling in the interval dy is

$$dp_2 = \frac{dy}{\sigma_2 \sqrt{(2\pi)}} \exp\left(-\frac{y^2}{2\sigma_2^2} \right),$$

and the compound probability of u falling in the interval du and, at the same time, y in the interval dy is the product $dp_1 dp_2$. Integrating with respect to y over its range of variation, we have the probability that u will fall in the interval du, irrespective of the value of y, as

$$dp = \frac{du}{2\pi\sigma_1\sigma_2} \int_{-\infty}^{\infty} \exp\left[-\frac{(u-y)^2}{2\sigma_1^2} - \frac{y^2}{2\sigma_2^2} \right] dy.$$

In this expression the argument of the exponential may be put in the form

$$-\frac{u^2}{2(\sigma_1^2+\sigma_2^2)}-\frac{\sigma_1^2+\sigma_2^2}{2\sigma_1^2\sigma_2^2}\left(y-\frac{u\sigma_2^2}{\sigma_1^2+\sigma_2^2}\right)^2.$$

If then we change the variable of integration from y to t, where

$$t=y-\frac{u\sigma_2^2}{\sigma_1^2+\sigma_2^2},$$

we have

$$dp=\frac{du}{2\pi\sigma_1\sigma_2}\exp\left[-\frac{u^2}{2(\sigma_1^2+\sigma_2^2)}\right]\int_{-\infty}^{\infty}\exp\left[-\frac{(\sigma_1^2+\sigma_2^2)\,t^2}{2\sigma_1^2\sigma_2^2}\right]dt$$

$$=\frac{du}{\sqrt{[2\pi(\sigma_1^2+\sigma_2^2)]}}\exp\left[-\frac{u^2}{2(\sigma_1^2+\sigma_2^2)}\right]$$

in virtue of (10). Thus u is normally distributed about zero as mean, with variance $\sigma_1^2+\sigma_2^2$. The reader should have no difficulty in removing the restriction of a common zero mean for the variates.

The following is still *another proof* of the theorem. With the same notation let

$$u=x+y,\quad v=-\frac{\sigma_2}{\sigma_1}x+\frac{\sigma_1}{\sigma_2}y.$$

Then

$$\frac{u^2+v^2}{2(\sigma_1^2+\sigma_2^2)}=\frac{x^2}{2\sigma_1^2}+\frac{y^2}{2\sigma_2^2},$$

and, as the values of x and y range from $-\infty$ to ∞, so do those of u and v. The probability that x and y fall simultaneously in the respective intervals dx and dy is

$$\frac{1}{2\pi\sigma_1\sigma_2}\exp\left[-\frac{1}{2}\left(\frac{x^2}{\sigma_1^2}+\frac{y^2}{\sigma_2^2}\right)\right]dx\,dy,$$

so that the probability density for the joint distribution of x and y, i.e. the probability per unit area at the point (x,y), is

$$\frac{1}{2\pi\sigma_1\sigma_2}\exp\left[-\frac{1}{2}\left(\frac{x^2}{\sigma_1^2}+\frac{y^2}{\sigma_2^2}\right)\right]=\frac{1}{2\pi\sigma_1\sigma_2}\exp\left[-\frac{1}{2}\left(\frac{u^2+v^2}{\sigma_1^2+\sigma_2^2}\right)\right].$$

Now the area of the element of the xy plane, bounded by the curves along which u has the values u and $u+du$ respectively, and those along which v has the values v and $v+dv$, is

$$\left| \frac{\partial(x,y)}{\partial(u,v)} \right| du\,dv = \frac{\sigma_1\sigma_2}{\sigma_1^2+\sigma_2^2} du\,dv,$$

where $\dfrac{\partial(x,y)}{\partial(u,v)}$ is the Jacobian of x and y with respect to u and v. Consequently the probability that, for a random choice of x and y, the representative point (x,y) will fall in this element of area is

$$\frac{1}{2\pi(\sigma_1^2+\sigma_2^2)} \exp\left[-\frac{1}{2}\left(\frac{u^2+v^2}{\sigma_1^2+\sigma_2^2}\right) \right] du\,dv$$
$$= \frac{du}{\sigma\sqrt{(2\pi)}} \exp\left(-\frac{u^2}{2\sigma^2}\right) \frac{dv}{\sigma\sqrt{(2\pi)}} \exp\left(-\frac{v^2}{2\sigma^2}\right),$$

where $\sigma^2 = \sigma_1^2+\sigma_2^2$. Since this is the probability that simultaneously u will lie in the interval du and v in the interval dv, it follows that these variates are independent, and that each is normally distributed with zero as mean and $\sigma_1^2+\sigma_2^2$ as variance.

9. Using the formulae of Ex. I, 11 show that, for the binomial distribution, the factorial moments about the mean are

$$\mu_{(2)} = npq, \quad \mu_{(3)} = -2npq(p+1),$$

and that, for the Poissonian distribution,

$$\mu_{(2)} = m, \quad \mu_{(3)} = -2m, \quad \mu_{(4)} = 3m(m+2).$$

MATHEMATICAL NOTES

NOTE I

Derivation of the Poissonian distribution (see §18)

In the binomial distribution the probability of the value r of the variate is $\binom{n}{r} p^r q^{n-r}$. We require the limiting value of this expression when $p = m/n$ and n tends to infinity, m being constant. The expression may be written

$$\frac{n!}{r!(n-r)!} \left(\frac{m}{n}\right)^r \left(1-\frac{m}{n}\right)^{n-r} = \frac{m^r}{r!} \left(1-\frac{m}{n}\right)^n \times \frac{n!}{(n-r)!\,n^r(1-m/n)^r}.$$

The limiting value of that part which precedes the sign of multiplication is $m^r e^{-m}/r!$. Then, using Stirling's formula for $n!$, viz.

$$n! \sim \sqrt{(2\pi n)}\, n^n e^{-n}$$

(the limiting value of the ratio of these two expressions, as n tends to infinity, being unity), we have

$$\lim \frac{n!}{(n-r)!\, n^r (1-m/n)^r}$$

$$= \lim \frac{\sqrt{(2\pi n)}\, n^n e^{-n}}{\sqrt{\{2\pi(n-r)\}}\,(n-r)^{n-r} e^{-(n-r)} n^r (1-m/n)^r}$$

$$= \lim \frac{1}{e^r (1-r/n)^{n-r+\frac{1}{2}}\,(1-m/n)^r} = \frac{1}{e^r e^{-r}} = 1,$$

since r is finite. Consequently

$$\lim \binom{n}{r} p^r q^{n-r} = \frac{m^r e^{-m}}{r!},$$

which is the required probability of the value $x = r$ in the Poissonian distribution with mean m.

NOTE II

Derivation of the normal distribution (see §19)

To examine the behaviour of the probability

$$P = \frac{n!\, p^{np+x} q^{nq-x}}{(np+x)!\,(nq-x)!}$$

as n tends to infinity, replace the factorials by their values given by Stirling's formula (Note I). Then an easy algebraical simplification leads to

$$P = \frac{1}{N\sqrt{(2\pi npq)}},$$

where

$$N = \left(1+\frac{x}{np}\right)^{np+x+\frac{1}{2}} \left(1-\frac{x}{nq}\right)^{nq-x+\frac{1}{2}}$$

Taking logarithms of both sides we may write, provided $|x|$ is less than the smaller of the two quantities np and nq,

$$\log N = (np + x + \tfrac{1}{2}) \left(\frac{x}{np} - \frac{x^2}{2n^2p^2} + \frac{x^3}{3n^3p^3} - \cdots \right)$$

$$- (nq - x + \tfrac{1}{2}) \left(\frac{x}{nq} + \frac{x^2}{2n^2q^2} + \frac{x^3}{3n^3q^3} + \cdots \right)$$

$$= \frac{x}{2n} \left(\frac{1}{p} - \frac{1}{q} \right) + \frac{x^2}{2npq} - \frac{x^2}{4n^2} \left(\frac{1}{p^2} + \frac{1}{q^2} \right) + \frac{x^3}{6n^2} \left(\frac{1}{q^2} - \frac{1}{p^2} \right)$$

$$+ \text{ terms with higher powers of } 1/n.$$

Introducing a new variable, $z = x/\sqrt{n}$, we may write the above

$$\log N = \frac{z}{2\sqrt{n}} \left(\frac{1}{p} - \frac{1}{q} \right) + \frac{z^2}{2pq} - \frac{z^2}{4n} \left(\frac{1}{p^2} + \frac{1}{q^2} \right) + \frac{z^3}{6\sqrt{n}} \left(\frac{1}{q^2} - \frac{1}{p^2} \right) + \cdots.$$

This series is convergent so long as $|z|$ is less than the smaller of the quantities $p\sqrt{n}$ and $q\sqrt{n}$. Now let n tend to infinity. Then, provided that neither p nor q is very small, and that $|z|$ is either finite or of lower order than $\sqrt[6]{n}$, all the terms of the above series tend to zero except $z^2/2pq$; so that, when n tends to infinity, we have

$$\log N = \frac{z^2}{2pq}$$

or
$$N = \exp(z^2/2pq).$$

Corresponding to unit increment in x we have the increment $1/\sqrt{n}$ in z, which may be denoted by dz when n tends to infinity. And, if we write dP for the limiting value of P, the above formula for P becomes

$$dP = \frac{dz}{\sqrt{(2\pi pq)}} \exp \left(- \frac{z^2}{2pq} \right),$$

giving the probability of z falling in the interval dz; and we have the required continuous distribution of z.

NOTE III

An important integral

To evaluate the integral

$$I = \int_0^\infty \exp(-x^2) \, dx,$$

let us write

$$I_1 = \int_0^a \exp(-x^2)\,dx = \int_0^a \exp(-y^2)\,dy.$$

Then

$$I_1^2 = \int_0^a \int_0^a \exp(-x^2 - y^2)\,dx\,dy.$$

If we regard x, y as Cartesian coordinates in a plane, the integration is extended over the area of the square $OACB$, bounded by $x = 0$, $x = a$, $y = 0$, $y = a$ (see Fig. 4). In polar coordinates the above relation is

$$I_1^2 = \iint \exp(-r^2)\, r\, dr\, d\theta,$$

the integration being extended over the same square Since the integrand is positive, the integral is intermediate in value between the integrals over the quadrants of circles, with centre O and radii a and $a\sqrt{2}$ respectively. Consequently I_1^2 lies in value between

$$\int_0^{\frac{1}{2}\pi} d\theta \int_0^a r \exp(-r^2)\,dr, \quad \text{and} \quad \int_0^{\frac{1}{2}\pi} d\theta \int_0^{a\sqrt{2}} r \exp(-r^2)\,dr.$$

But, as a tends to infinity, each of these integrals converges to $\frac{1}{4}\pi$. Hence

$$I^2 = \lim I_1^2 = \tfrac{1}{4}\pi,$$

and therefore

$$\int_0^\infty \exp(-x^2)\,dx = \tfrac{1}{2}\sqrt{\pi}.$$

The substitution $x = z/\sigma\sqrt{2}$ then gives

$$\int_0^\infty \exp\left(-\frac{z^2}{2\sigma^2}\right)dz = \sigma\sqrt{\frac{\pi}{2}},$$

and, since the integrand is an even function of z,

$$\int_{-\infty}^\infty \exp\left(-\frac{z^2}{2\sigma^2}\right)dz = \sigma\sqrt{(2\pi)}.$$

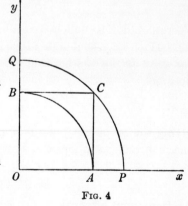

Fig. 4

CHAPTER IV

BIVARIATE DISTRIBUTIONS.
REGRESSION AND CORRELATION

23. Discrete distributions. Moments

Suppose that we have records for N marriages, giving the ages of bridegroom and bride, x and y years respectively. Then to each marriage corresponds a pair of values (x_i, y_i) of the variables. Each pair may be represented by a point P_i in the x, y plane, the coordinates of the point being the pair of values represented. Such a graphical representation is called a *scatter diagram*. Possibly the pairs of values are not all different. If the pair (x_i, y_i) occurs f_i times, then f_i is the frequency of that pair; and, as in the case of a single variable,

$$\sum_{i=1}^{n} f_i = N, \tag{1}$$

n being the number of different pairs of values. The assemblage of pairs of values, together with their frequencies, constitutes a bivariate frequency distribution. Other examples of bivariate distributions are furnished by the heights and the weights of a group of men, or by the amount of fertilizer per acre and the yield of grain per acre on a number of different plots of land.

The moments of a bivariate distribution are generalizations of those of a univariate one. Thus the moment μ'_{rs} about the origin $(0, 0)$, of order r in x and s in y, is defined by

$$\mu'_{rs} = \frac{1}{N} \sum_i f_i x_i^r y_i^s. \tag{2}$$

In particular

$$\mu'_{10} = \frac{1}{N} \sum f_i x_i = \bar{x}, \quad \mu'_{01} = \frac{1}{N} \sum f_i y_i = \bar{y} \tag{3}$$

where $\bar{x}$ is the mean value of x in the distribution, and $\bar{y}$ the mean value of y. Similarly,

$$\mu'_{20} = \frac{1}{N} \sum f_i x_i^2 = \sigma_x^2 + \bar{x}^2, \quad \mu'_{02} = \frac{1}{N} \sum f_i y_i^2 = \sigma_y^2 + \bar{y}^2, \tag{4}$$

where σ_x^2 is the variance of the variable x in the distribution, or the moment μ_{20} about the mean $(\bar{x}, \bar{y})$; and likewise σ_y^2 is the variance of y, or the moment μ_{02} about the mean. Lastly, we may mention the moment μ_{11}', which is given by

$$\mu_{11}' = \frac{1}{N} \sum_i f_i x_i y_i.$$

The corresponding moment μ_{11} about the mean of the distribution is called the *covariance** of the variables. Thus

$$\begin{aligned} \mu_{11} &= \frac{1}{N} \sum_i f_i (x_i - \bar{x})(y_i - \bar{y}) \\ &= \mu_{11}' - \bar{x} \sum f_i y_i / N - \bar{y} \sum f_i x_i / N + \overline{xy} \\ &= \mu_{11}' - \overline{xy} \end{aligned} \tag{5}$$

in virtue of (1) and (3). This formula is very important, and will be used frequently.

24. Continuous distributions

A continuous bivariate distribution includes all pairs of values represented by points within a certain region of the xy plane, each pair of values occurring at least once. The number of pairs of values is therefore infinite. Distributions of the type ordinarily employed have each for *relative frequency density* a continuous function $f(x, y)$, which is such that the relative frequency for the infinitesimal rectangular region, of area $dx\,dy$, and defined by

$$x - \tfrac{1}{2}dx \leqslant x \leqslant x + \tfrac{1}{2}dx, \quad y - \tfrac{1}{2}dy \leqslant y \leqslant y + \tfrac{1}{2}dy,$$

has the value $f(x, y)\,dx\,dy$. Thus $f(x, y)$ is the relative frequency per unit area at the point (x, y), and the sum of the relative frequencies for all elements of the distribution is unity, so that

$$\iint f(x, y)\,dx\,dy = 1, \tag{6}$$

---

* Many writers denote this quantity by p. We are, however, loth to overwork the symbol p by giving it still another meaning.

the integration extending over the region* of the plane which represents the distribution. The mean $(\bar{x}, \bar{y})$ of the distribution is given by

$$\bar{x} = \iint x f(x, y)\,dx\,dy, \quad \bar{y} = \iint y f(x, y)\,dx\,dy. \tag{7}$$

For higher moments we have formulae corresponding to those of a discrete distribution. Thus

$$\mu_{20} = \iint (x - \bar{x})^2 f(x, y)\,dx\,dy$$
$$= \mu'_{20} - 2\bar{x} \iint x f(x, y)\,dx\,dy + \bar{x}^2 = \mu'_{20} - \bar{x}^2$$

which is equivalent to

$$\sigma_x^2 = \mu'_{20} - \bar{x}^2, \tag{8}$$

and similarly, $\qquad\qquad \sigma_y^2 = \mu'_{02} - \bar{y}^2. \tag{9}$

The covariance, being the first product moment about the mean, is

$$\mu_{11} = \iint (x - \bar{x})(y - \bar{y}) f(x, y)\,dx\,dy$$
$$= \mu'_{11} - \bar{x} \iint y f(x, y)\,dx\,dy - \bar{y} \iint x f(x, y)\,dx\,dy + \bar{x}\bar{y}$$
$$= \mu'_{11} - \bar{x}\bar{y}, \tag{10}$$

in virtue of (6) and (7).

It will be sufficient, as a rule, to give proofs of theorems for a discrete distribution. The student can easily rewrite these for the case of a continuous distribution, replacing sums by definite integrals, and the relative frequency f_i/N by $f(x, y)\,dx\,dy$. Some of the steps will be indicated in Examples.

25. Lines of regression

It frequently happens that the scatter diagram indicates an association between the variables, x and y, the distribution of dots being denser in the neighbourhood of a certain curve, which may be called a *curve of regression*. The equation of such a curve indicates a functional relationship to which the association of the variables

---

* By defining $f(x, y)$ as equal to zero outside this region, we may make the range of integration $-\infty$ to ∞ for each variable.

approximates, more or less roughly. We shall consider later the problem of fitting different curves to the data, so as to obtain the curve of a specified form which best fits the data. For the present we confine our attention to the straight line which is the best fit, or more definitely, to the problem of determining two straight lines, one of which gives the closest estimate a straight line can give to the average value of y for each specified value of x, while the other gives the corresponding estimate of x for a given value of y. These are called the *lines of regression* of y on x, and of x on y respectively.

Consider first the line of regression of y on x. We have to determine constants a and b so that the equation

$$y = a + bx \qquad (11)$$

gives, for each value of x, the best estimate a linear equation can give for the average value of y. We interpret the term 'best estimate' in accordance with the principle of least squares; that is to say, we find a and b so as to minimize the sum of the squares of the deviations of the actual values of y in the distribution from their estimates given by (11). Thus if P_i is the point of the diagram representing the pair of values (x_i, y_i), and H_i is the point on the straight line (11)

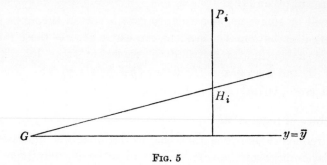

Fig. 5

with the same abscissa x_i, the deviation of P_i from its estimate is $H_i P_i$, whose value is clearly $y_i - (a + bx_i)$. Thus we have to choose a and b so as to make $\sum_i f_i(y_i - a - bx_i)^2$ a minimum, f_i being the frequency of the pair of values (x_i, y_i). For a minimum value the partial derivatives of the above expression with respect to a and b must both be zero. Hence the equations for determining these constants are

$$\Sigma f_i(y_i - a - bx_i) = 0, \quad \Sigma f_i x_i(y_i - a - bx_i) = 0, \tag{12}$$

which are called the *normal equations*. In virtue of (3), (4) and (5) they are equivalent to

$$\bar{y} - a - b\bar{x} = 0 \tag{13}$$

and

$$\mu_{11} + \bar{x}\bar{y} - a\bar{x} - b(\sigma_x^2 + \bar{x}^2) = 0. \tag{14}$$

The first of these shows that the required line passes through the mean $(\bar{x}, \bar{y})$ of the distribution. Also, on eliminating a between (13) and (14), we find

$$\mu_{11} - b\sigma_x^2 = 0.$$

Consequently the gradient b of the line of regression of y on x is given by

$$b = \mu_{11}/\sigma_x^2, \tag{15}$$

and, since the line passes through $(\bar{x}, \bar{y})$, its equation may be expressed

$$y - \bar{y} = \frac{\mu_{11}}{\sigma_x^2}(x - \bar{x}). \tag{16}$$

If the mean of the distribution is taken as origin, the line of regression of y on x is

$$y = \mu_{11}x/\sigma_x^2. \tag{17}$$

The gradient μ_{11}/σ_x^2 is often called the *coefficient of regression* of y on x.

Similarly, or by interchanging variables, we find that the line of regression of x on y is

$$x - \bar{x} = \frac{\mu_{11}}{\sigma_y^2}(y - \bar{y}), \tag{18}$$

and μ_{11}/σ_y^2 is the coefficient of regression of x on y. The product of the two coefficients of regression is symmetrical with respect to x and y. Its square root, $\mu_{11}/\sigma_x\sigma_y$, is the *coefficient of correlation*, r. Thus

$$r = \frac{\mu_{11}}{\sigma_x\sigma_y}, \tag{19}$$

having the same sign as the covariance μ_{11}.

Example 1. Find the tangent of the inclination, β, of (18) to (16). Since the gradients of the two lines are σ_y^2/μ_{11} and μ_{11}/σ_x^2, we have

$$\tan\beta = \left(\frac{\sigma_y^2}{\mu_{11}} - \frac{\mu_{11}}{\sigma_x^2}\right) \div \left(1 + \frac{\sigma_y^2}{\sigma_x^2}\right)$$

$$= \frac{1-r^2}{r}\frac{\sigma_x\sigma_y}{\sigma_x^2+\sigma_y^2}.$$

Example 2. For a continuous bivariate distribution the mean square deviation from the line of regression of y on x is

$$\iint (y-a-bx)^2 f(x,y)\,dx\,dy.$$

By minimizing this for variation of a and b, show that the normal equations are

$$\iint (y-a-bx)f(x,y)\,dx\,dy = 0 = \iint x(y-a-bx)f(x,y)\,dx\,dy,$$

and that these are expressible in the forms (13) and (14).

26. Coefficient of correlation. Standard error of estimate

The significance of the correlation coefficient, r, as a measure of the closeness of the association of the variables x and y, will be apparent from the theorem now to be considered. The equation of the line of regression of y on x was found by minimizing the sum of the squares of the deviations $H_i P_i$. We shall now prove that the sum of the squares of these deviations from the line of regression of y on x is equal to $N\sigma_y^2(1-r^2)$.

Let the mean of the distribution be taken as origin, so that $\bar{x} = \bar{y} = 0$. Then, in virtue of (17), $H_i P_i$ is equal to $y_i - bx_i$; and the required sum of squares of the deviations is

$$\begin{aligned}
\sum_i f_i(y_i - bx_i)^2 &= \sum f_i y_i^2 - 2b\sum f_i x_i y_i + b^2 \sum f_i x_i^2 \\
&= N\sigma_y^2 - 2Nb\mu_{11} + Nb^2\sigma_x^2 \\
&= N\sigma_y^2\left(1 - \frac{\mu_{11}^2}{\sigma_x^2\sigma_y^2}\right) \\
&= N\sigma_y^2(1-r^2), \tag{20}
\end{aligned}$$

as stated above. Denoting this sum of squares by NS_y^2 we have

$$S_y^2 = \sigma_y^2(1-r^2) \tag{21}$$

or

$$r^2 = 1 - \frac{S_y^2}{\sigma_y^2}. \tag{22}$$

Since S_y^2 is the mean square deviation of points from the line of regression of y on x, S_y is called the *standard error of estimate* of y from the regression equation (16). In the same way the sum of the squares of the deviations of points from the line of regression of x on y, measured parallel to the x-axis, is NS_x^2, where

$$S_x^2 = \sigma_x^2(1 - r^2), \tag{23}$$

and S_x is the standard error of estimate of x from the regression equation (18).

Since the sum of squares of deviation cannot be negative, it follows from (20) that $r^2 \leqslant 1$, or

$$-1 \leqslant r \leqslant 1. \tag{24}$$

If $r = 1$ or -1, the sum of squares of deviations from either line of regression is zero. Consequently each deviation is zero, and all the points lie on both lines of regression. These two lines then coincide, and there is a linear functional relation between the variables x and y, giving perfect correlation. The nearer r^2 is to unity, the closer are the points to the lines of regression, and the nearer are these two lines to coincidence (cf. § 25, Ex. 1). Thus *the magnitude of r may be taken as a measure of the degree to which the association between the variables approaches a linear functional relationship.* The sign of r is the same as that of the covariance μ_{11}, and therefore also the same as that of the gradients of the lines of regression. Hence r is positive when, on the whole, y increases with x, and negative when y decreases as x increases. When r is zero the variables are usually described as *uncorrelated.*

The coefficient of correlation between two variables is also the coefficient of correlation between the deviations of the variables from their means. For the value of r depends only on μ_{11}, σ_x and σ_y, and these are functions of the above deviations. Thus

$$r = \frac{\sum f_i(x_i - \bar{x})(y_i - \bar{y})}{\sqrt{[\sum f_i(x_i - \bar{x})^2]}\sqrt{[\sum f_i(y_i - \bar{y})^2]}},$$

and, in virtue of (3), (4) and (5), this may be expressed

$$r = \frac{\sum f_i x_i y_i - (\sum f_i x_i)(\sum f_i y_i)/N}{\sqrt{[\sum f_i x_i^2 - (\sum f_i x_i)^2/N]}\sqrt{[\sum f_i y_i^2 - (\sum f_i y_i)^2/N]}}. \tag{25}$$

This formula is sometimes convenient.

Example. Taking the mean as origin for a continuous distribution, show that the mean square deviation, S_y^2, from the line of regression of y on x is given by

$$S_y^2 = \int\int (y-bx)^2 f(x,y)\,dx\,dy$$
$$= \sigma_y^2 - 2b\mu_{11} + b^2\sigma_x^2 = \sigma_y^2(1-r^2).$$

27. Estimates from the regression equation

A few simple relations between the y_i and their estimates, Y_i, given by the regression equation (11) or (16), play an important part in later work. Thus

$$Y_i = a + bx_i, \tag{26}$$

and the normal equations (12) are expressible as

$$\Sigma f_i(y_i - Y_i) = 0 \tag{27}$$

and $\quad\quad\quad\Sigma f_i x_i(y_i - Y_i) = 0. \tag{28}$

From (27) it follows that the mean of the Y's is equal to the mean $\bar{y}$ of the y's. Also, on multiplying (27) and (28) by a and b respectively and adding, we deduce

$$\Sigma f_i Y_i(y_i - Y_i) = 0, \tag{29}$$

and therefore, in virtue of (27),

$$\Sigma f_i(y_i - Y_i)(Y_i - \bar{y}) = 0. \tag{30}$$

This relation is very important.

The sum of the squares of the deviations of the y's from their mean may then be expressed

$$\Sigma f_i(y_i - \bar{y})^2 = \Sigma f_i[(y_i - Y_i) + (Y_i - \bar{y})]^2$$
$$= \Sigma f_i(y_i - Y_i)^2 + \Sigma f_i(Y_i - \bar{y})^2, \tag{31}$$

since the sum of the products vanishes by (30). Now the first member of (31) has the value $N\sigma_y^2$. The first sum in the second member has been proved equal to $N\sigma_y^2(1-r^2)$. Consequently

$$\Sigma f_i(Y_i - \bar{y})^2 = Nr^2\sigma_y^2, \tag{32}$$

showing that the variance of the Y's is r^2 times that of the y's; or

$$\sigma_Y^2 = r^2\sigma_y^2, \quad \sigma_Y = |r|\,\sigma_y. \tag{33}$$

Finally, we may show that the coefficient of correlation between the y_i and their estimates Y_i is equal to $|r|$. Take the origin at the common mean of these variables, so that $\bar{y} = 0$. Then, in virtue of (29),

$$\Sigma f_i y_i Y_i = \Sigma f_i Y_i^2 = N\sigma_Y^2,$$

and the coefficient of correlation between the y's and the Y's is

$$\frac{\Sigma f_i y_i Y_i / N}{\sigma_y \sigma_Y} = \frac{\sigma_Y^2}{\sigma_y \sigma_Y} = \frac{\sigma_Y}{\sigma_y} = |r|. \tag{34}$$

Example 1. Prove (34) as follows. Taking the mean of the distribution as origin, so that $Y_i = bx_i$, $\sigma_Y = |b|\,\sigma_x$, we have for the required correlation coefficient

$$\frac{\Sigma f_i y_i Y_i}{N\sigma_y \sigma_Y} = \frac{b\Sigma f_i x_i y_i}{N |b| \sigma_x \sigma_y} = \frac{br}{|b|} = |r|,$$

since r and b have the same sign.

Example 2. Show that the normal equations for a continuous distribution (cf. § 25, Ex. 2) are expressible as

$$\int\int (y - Y)f(x, y)\,dx\,dy = 0, \quad \int\int x(y - Y)f(x, y)\,dx\,dy = 0,$$

and from these deduce the relations

$$\int\int Y(y - Y)f(x, y)\,dx\,dy = 0, \quad \int\int (y - Y)(Y - \bar{y})f(x, y)\,dx\,dy = 0$$

corresponding to (29) and (30). Hence show that

$$\sigma_y^2 = \int\int (y - Y)^2 f(x, y)\,dx\,dy + \int\int (Y - \bar{y})^2 f(x, y)\,dx\,dy,$$

and deduce that $\qquad \int\int (Y - \bar{y})^2 f(x, y)\,dx\,dy = r^2 \sigma_y^2.$

28. Change of units

Before illustrating the above theory by a numerical example, let us examine the effect of a change of units on the calculation of r, μ_{11} and b. As in § 2, let u be the measure of the deviation of x from an origin $x = a$, in terms of a unit c times the original x-unit, so that $x = a + cu$. Then, if x' and u' are the deviations of the two variables from their means, $x' = cu'$, and

$$\sigma_x^2 = \frac{1}{N} \Sigma f_i x_i'^2 = \frac{c^2}{N} \Sigma f_i u_i'^2 = c^2 \sigma_u^2.$$

Similarly, if v is the measure of the deviation of y from an origin $y = a'$, in terms of a unit c' times the original y-unit, $y = a' + c'v$, $y' = c'v'$ and $\sigma_y = c'\sigma_v$. Then the coefficient of correlation between x and y is

$$r = \frac{\sum f_i x_i' y_i'/N}{\sigma_x \sigma_y} = \frac{cc' \sum f_i u_i' v_i'/N}{cc' \sigma_u \sigma_v} = \frac{\text{covariance of } u \text{ and } v}{\sigma_u \sigma_v},$$

and is thus equal to the correlation between u and v. The value of r is therefore independent of the units employed, and is thus an absolute measure of correlation. But the covariance of x and y is

$$\mu_{11} = \sum f_i x_i' y_i'/N = cc' \sum f_i u_i' v_i'/N$$
$$= cc' \text{ (covariance of } u \text{ and } v\text{).} \qquad (35)$$

Similarly, the regression coefficient of y on x is

$$b = \frac{\mu_{11}}{\sigma_x^2} = \frac{c'}{c} \text{ (regression coefficient of } v \text{ on } u\text{).} \qquad (36)$$

If then c and c' are equal, the value b is the same for both pairs of variables.

29. Numerical illustration

For 1,000 marriages the ages of bridegroom and bride, x and y years respectively, are grouped in the table below with class interval of 5 years for each, the frequencies for the different classes being shown in the body of the table. Find the regression equations and the coefficient of correlation between the variables.

Such a table is called a correlation table. The values of x and y indicated are the mid-values in the classes. Thus for the class in which the age of the bridegroom is between 25 and 30, and the age of the bride between 20 and 25, the values of x and y are taken as $27 \cdot 5$ and $22 \cdot 5$ respectively, and the frequency is 190. The data represented in any one column constitute a vertical *array*, or an array of y's, because in each such array y assumes different values while x remains constant. Similarly, the data in any row constitute a horizontal array, or an array of x's. In the row prefixed N_c the frequencies are given for the individual columns, and in the column headed N_r the frequencies for the separate rows.

Let us take as new origin the point $(27 \cdot 5, 27 \cdot 5)$, whose coordinates are the mid-values of the class 25 to 30 for x and y; and, as a new unit for each variable, the common class interval of 5 years. Then the deviations, u and v, of the ages of bridegroom and bride from the new origin in terms of the new unit are those shown in the second row and the second column. The

Ages of Bridegroom and Bride

y \\ x	v \\ u	17·5 (−2)	22·5 (−1)	27·5 (0)	32·5 (1)	37·5 (2)	42·5 (3)	47·5 (4)	52·5 (5)	57·5 (6)	N_r	vN_r	v^2N_r	U	vU
17·5	−2	11	62	19	3	1	—	—	—	—	96	−192	384	−79	158
22·5	−1	6	220	190	34	6	2	—	—	—	458	−458	458	−180	180
27·5	0	—	46	165	59	13	5	2	1	—	291	0	0	67	0
32·5	1	—	3	25	33	14	6	3	2	—	86	86	86	98	98
37·5	2	—	1	3	8	9	6	4	3	1	35	70	140	80	160
42·5	3	—	—	—	1	3	5	4	3	2	18	54	162	65	195
47·5	4	—	—	—	—	1	2	3	3	2	11	44	176	47	188
52·5	5	—	—	—	—	—	—	1	2	2	5	25	125	26	130
N_c		17	332	402	138	47	26	17	14	7	1,000	−371	1,531	124	1,109
uN_c		−34	−332	0	138	94	78	68	70	42	124				
u^2N_c		68	332	0	138	188	234	272	350	252	1,834				
V		−28	−339	−197	12	37	39	40	39	26	−371				
uV		56	339	0	12	74	117	160	195	156	1,109				

fourth row from the bottom of the table, prefixed uN_c, gives the sum Σfu for each column; and these are added horizontally to give the total sum 124 for the distribution. The next row gives the sum Σfu^2 for each column, with total sum 1,834 for the distribution. The row prefixed V gives for each column the sum Σfv. Thus the sum -28, in the column $u = -2$, is obtained from

$$11(-2) + 6(-1) = -28.$$

The last row gives the sum Σfuv for each column, each entry being obtained by multiplying the value of V above by the common value of u for that column. Summing horizontally all the values uV, we have the product sum Σfuv for the whole distribution, viz. 1,109.

The columns to the right of the table are explained similarly. That headed U gives for each row the sum Σfu. Thus the first entry -79 in the column is obtained from

$$11(-2) + 62(-1) + 19 \times 0 + 3 \times 1 + 1 \times 2 = -79.$$

The last column, headed vU, gives the sum Σfuv for each row, any entry being obtained by multiplying the value of U to the left by the common value of v for that row. Summing the entries in this column we find again the product sum Σfuv for the whole distribution, thus providing a check on the calculations.

Using the values thus obtained we have

$$\bar{u} = 124/1000 = 0 \cdot 124, \quad \bar{v} = -371/1000 = -0 \cdot 371.$$

Therefore

$$\bar{x} = 27 \cdot 5 + 5(0 \cdot 124) = 28 \cdot 120, \quad \bar{y} = 27 \cdot 5 - 5(0 \cdot 371) = 25 \cdot 645,$$

giving the mean ages of bridegroom and bride. The variance of u is

$$\sigma_u^2 = 1 \cdot 834 - (0 \cdot 124)^2 = 1 \cdot 8186,$$

so that

$$\sigma_u = 1 \cdot 349 = 1 \cdot 35 \quad \text{nearly,}$$

and therefore

$$\sigma_x = 6 \cdot 745 = 6 \cdot 75 \quad \text{nearly.}$$

Similarly, we find

$$\sigma_v^2 = 1 \cdot 531 - (0 \cdot 371)^2 = 1 \cdot 3934,$$

$$\sigma_v = 1 \cdot 18; \quad \sigma_y = 5 \cdot 90.$$

The coefficient of correlation between x and y is equal to that between u and v. Thus

$$r = \frac{\Sigma uv/N - \bar{u}\bar{v}}{\sigma_u \sigma_v} = \frac{1 \cdot 109 + 0 \cdot 371 \times 0 \cdot 124}{1 \cdot 349 \times 1 \cdot 180}$$

$$= 0 \cdot 726 = 0 \cdot 73 \quad \text{nearly.}$$

Since the units for u and v are equal, the regression coefficient of y on x is equal to that of v on u, which is

$$\text{(covariance of } u \text{ and } v)/\sigma_u^2 = 1 \cdot 155/1 \cdot 8186 = 0 \cdot 635.$$

The line of regression of y on x is therefore

$$y - 25 \cdot 645 = 0 \cdot 635(x - 28 \cdot 12),$$

that is
$$y = 7 \cdot 79 + 0 \cdot 635x.$$

The student may find similarly that the line of regression of x on y is

$$x = 6 \cdot 86 + 0 \cdot 829y.$$

30. Correlation of ranks

A group of n individuals may be arranged in order of merit or proficiency in the possession of a certain characteristic. The same group would, as a rule, give different orders for different characteristics. Considering the orders corresponding to two characteristics, A and B, let x_i, y_i be the ranks of the ith individual in A and B respectively. Then the coefficient of correlation between the x's and the y's is called the *rank correlation* coefficient in the characteristics A and B for that group of individuals. On the assumption that no two individuals are bracketed equal in either classification, each of the variables takes the values 1, 2, 3, ..., n; and therefore

$$\bar{x} = \tfrac{1}{2}(n+1) = \bar{y}. \tag{37}$$

As a rule x_i is not equal to y_i. Let d_i denote the difference, so that

$$d_i = x_i - y_i. \tag{38}$$

Then, if x' and y' denote the deviations of the variables from their means, we have also

$$d_i = x'_i - y'_i. \tag{39}$$

The coefficient of correlation between the variables is given by

$$r = \frac{\sum x'_i y'_i}{\sqrt{[(\sum x_i'^2)(\sum y_i'^2)]}}. \tag{40}$$

To express this in terms of n and the differences, d_i, we observe that the variance of each of the variables x and y is $(n^2-1)/12$ (cf. § 3, Ex. 2). Therefore

$$\sum x_i'^2 = \sum y_i'^2 = (n^3 - n)/12. \tag{41}$$

Also $\quad \sum d_i^2 = \sum (x'_i - y'_i)^2 = \sum x_i'^2 + \sum y_i'^2 - 2\sum x'_i y'_i,$

and thus, in virtue of (41),

$$\sum x'_i y'_i = \tfrac{1}{12}(n^3 - n) - \tfrac{1}{2}\sum d_i^2.$$

Substitution of these values in (40) gives

$$r = 1 - \frac{6 \sum d_i^2}{n^3 - n}. \tag{42}$$

This is the required formula for the coefficient of correlation of ranks.

If the correlation is perfect all the d's are zero, and $r = 1$. If the orders in the two characteristics are exactly the reverse of each other, $x_i + y_i = n + 1$. All the points of the scatter diagram then lie on a straight line with negative gradient. Consequently $r = -1$, and there is perfect inverse correlation.

Example. The ranks of the same 16 students in Mathematics and Physics were as follows, two numbers within brackets denoting the ranks of the same student in Mathematics and Physics respectively: $(1, 1), (2, 10), (3, 3), (4, 4),$ $(5, 5), (6, 7), (7, 2), (8, 6), (9, 8), (10, 11), (11, 15), (12, 9), (13, 14), (14, 12),$ $(15, 16), (16, 13)$. Calculate the rank correlation coefficient for proficiencies of this group in Mathematics and Physics.

Here d_i is the difference between the two numbers in the ith pair of brackets. It is easily verified that $\Sigma d_i^2 = 136$, $n^3 - n = 16 \times 255$. Consequently

$$r = 1 - \frac{6 \times 136}{16 \times 255} = 1 - \tfrac{1}{5} = 0 \cdot 8.$$

31. Bivariate probability distributions

The theorems proved for a bivariate frequency distribution in §§ 23–28 hold equally for a probability distribution of two variates, relative frequency in the former case being replaced by probability in the latter. Thus, for a discrete distribution, if p_i is the probability of the occurrence of the pair of values (x_i, y_i), the moment μ'_{rs} about the origin is

$$\mu'_{rs} = \sum_i p_i x_i^r y_i^s,$$

which is the expected value of the product $x^r y^s$. In particular the expected values of x and y are

$$E(x) = \sum p_i x_i, \quad E(y) = \sum p_i y_i,$$

while, corresponding to (4) and (5), we have

$$\sigma_x^2 = \mu'_{20} - [E(x)]^2, \quad \sigma_y^2 = \mu'_{02} - [E(y)]^2$$

and
$$\mu_{11} = \mu'_{11} - E(x)\, E(y),$$

μ_{11} being the covariance of the variates, which is the expected value of the product of their deviations from their means. The coefficient of correlation is defined by (19), which may here be expressed in the alternative form

$$r = \frac{E(x'y')}{\sigma_x \sigma_y} = \frac{E(x'y')}{\sqrt{[E(x'^2) E(y'^2)]}}, \tag{19'}$$

x', y' being the deviations of the variates from their expected values.

In the case of continuous probability distributions we confine our attention to those in which the probability that the variates will fall simultaneously in the intervals dx and dy is expressible in the form $\phi(x, y)\, dx\, dy$, the probability density $\phi(x, y)$ being continuous and essentially positive. Then we have formulae corresponding to (6)–(10), with $\phi(x, y)$ in place of $f(x, y)$.

The variates are *independent* if the probability distribution of each is independent of the value assumed by the other. In particular, if the variates are continuous, with probability densities $\phi_1(x)$ and $\phi_2(y)$ respectively, the probability that x will fall in the interval dx, and at the same time y will fall in the interval dy, is $\phi_1(x)\, dx\, \phi_2(y)\, dy$ by the theorem of compound probability. Then the probability density $\phi(x, y)$ for the bivariate distribution is of the form $\phi_1(x)\, \phi_2(y)$. Conversely, when this relation holds, the continuous variates are independent. Now it was proved in §§ 10 and 12 that the covariance of two independent variates is equal to zero. It follows from the above definition of r that *the coefficient of correlation of two independent variates is equal to zero*. The converse of this theorem, however, is not necessarily true; that is to say, uncorrelated variables are not necessarily independent.

The moment generating function of the bivariate distribution is defined as the expected value of the function $\exp(t_1 x + t_2 y)$, where t_1 and t_2 are independent of x and y. Thus, in the case of a continuous distribution, the m.g.f. with respect to the origin is

$$M(t_1, t_2) = \iint \exp(t_1 x + t_2 y)\, \phi(x, y)\, dx\, dy.$$

When this integral has a meaning the exponential may be expanded in powers of t_1 and t_2; and the coefficient of $t_1^r t_2^s / r! s!$ is the moment μ'_{rs} about the origin. In Ex. 3 at the end of Chapter v we shall find

D

this function for the bivariate normal distribution, and use it to calculate the moments.

When the variates x and y are independent we have

$$M(t_1, t_2) = E[\exp(t_1 x + t_2 y)] = E[\exp(t_1 x)] \cdot E[\exp(t_2 y)]$$
$$= (\text{function of } t_1) \times (\text{function of } t_2).$$

And conversely, when the m.g.f. is of this form, the variates are independent.

32. Variance of a sum of variates

Consider the variates x and y, with standard deviations σ_x and σ_y respectively. Their distributions determine a bivariate probability distribution for the two variates, with correlation coefficient r given by (19′); and each value that may be taken by their sum

$$u = x + y \tag{43}$$

has a definite probability. Then, in virtue of § 10 (10),

$$E(u) = E(x) + E(y), \tag{44}$$

and therefore, if x', y', u' are the deviations of the variates from their means, we obtain from (43) and (44) by subtraction

$$u' = x' + y'.$$

Consequently $\qquad u'^2 = x'^2 + y'^2 + 2x'y',$

and, on taking the expected value of each member we deduce, in virtue of (19′),

$$\sigma_u^2 = \sigma_x^2 + \sigma_y^2 + 2r\sigma_x \sigma_y. \tag{45}$$

This is the required formula for the variance of the sum of the variates. And, since $-r$ is the correlation between x and $-y$, it follows that the variance of the difference

$$v = x - y$$

is given by $\qquad \sigma_v^2 = \sigma_x^2 + \sigma_y^2 - 2r\sigma_x \sigma_y. \tag{46}$

If r is zero, as in the case of independence of x and y, we have the simple result

$$\sigma_u^2 = \sigma_v^2 = \sigma_x^2 + \sigma_y^2, \tag{47}$$

already proved in §§ 10, 15 and 16.

More generally, let u be a *linear function* of the variates $x, y, z, ...,$ so that

$$u = ax + by + cz + ..., \tag{48}$$

the constants $a, b, c, ...$ being either positive or negative. Then

$$E(u) = aE(x) + bE(y) + cE(z) + ...$$

and therefore by subtraction

$$u' = ax' + by' + cz' + ...,$$

the primes indicating that the variates are measured from their means. On squaring both sides, and taking expected values, we obtain the required formula

$$\sigma_u^2 = a^2\sigma_x^2 + b^2\sigma_y^2 + c^2\sigma_z^2 + ... + 2abr_{xy}\sigma_x\sigma_y + ..., \tag{49}$$

in which r_{xy} is the coefficient of correlation between x and y.

COLLATERAL READING

EZEKIEL, 1930, 2, chapters III–V and VIII–IX.
YULE and KENDALL, 1937, 1, chapter XI.
KENNEY, 1939, 3, part I, chapter VIII.
CAMP, 1934, 1, part I, chapters VIII and IX.
JONES, 1924, 3, chapters X and XI.
RIETZ, 1927, 2, pp. 77–88.
RIETZ (ed.), 1924, 1, chapter VIII (to p. 129).
RIDER, 1939, 6, §§18, 19, 22–24.
BOWLEY, 1920, 2, part II, chapters VI and VII.
PLUMMER, 1940, 1, chapter V.
MILLS, 1938, 1, chapter X.
GOULDEN, 1939, 2, chapters VI and VII.
SNEDECOR, 1938, 3, chapters VI and VII.
TIPPETT, 1931, 2, chapter VII.

EXAMPLES IV

1. Show that, if a and b are constants and r is the correlation between x and y, then the correlation between ax and by is equal to r if the signs of a and b are alike, and to $-r$ if they are different.

Also show that, if the constants a, b, c are positive, the correlation between $(ax + by)$ and cy is equal to

$$(ar\sigma_x + b\sigma_y)/\sqrt{(a^2\sigma_x^2 + b^2\sigma_y^2 + 2abr\sigma_x\sigma_y)}.$$

2. The variables x and y are connected by the equation $ax + by + c = 0$. Show that the correlation between them is -1 if the signs of a and b are alike, and $+1$ if they are different.

3. Show that, if x', y' are the deviations of the variables from their means,

$$r = 1 - \frac{1}{2N} \sum_i (x_i'/\sigma_x - y_i'/\sigma_y)^2$$

and

$$r = -1 + \frac{1}{2N} \sum_i (x_i'/\sigma_x + y_i'/\sigma_y)^2,$$

and deduce that $-1 \leqslant r \leqslant 1$. (Rietz, 1927, 2, p. 84.)

4. The variates x and y have zero means, the same variance σ^2 and zero correlation. Show that

$$(x \cos \alpha + y \sin \alpha) \quad \text{and} \quad (x \sin \alpha - y \cos \alpha)$$

have the same variance σ^2 and zero correlation.

5. *Weighted mean with minimum variance.* Let x_i $(i = 1, ..., n)$ be n independent variates with variances σ_i^2. If the variates are given weights w_i, their weighted mean is

$$\sum_i w_i x_i / \sum_i w_i = \sum_i c_i x_i,$$

where

$$c_i = w_i / \sum_i w_i.$$

We can show that the variance of this weighted mean is least when the weights are inversely proportional to variances σ_i^2. In virtue of (49) the variance of the weighted mean is

$$\sigma^2 = \sum_i c_i^2 \sigma_i^2 = (\sum_i w_i^2 \sigma_i^2) / (\sum_j w_j)^2.$$

For a minimum σ^2 the partial derivatives of this with respect to the w_i must be zero. This requires

$$(\sum w_i)^2 w_i \sigma_i^2 - (\sum w_i^2 \sigma_i^2)(\sum w_i) = 0,$$

so that

$$w_i \sigma_i^2 = (\sum w_i^2 \sigma_i^2) / (\sum w_i),$$

showing that $w_i \sigma_i^2$ is the same for all values of i. Thus the weights of the variates are inversely proportional to their variances.

Show that this minimum variance is equal to H/n, where H is the harmonic mean of the variances σ_i^2.

6. For a given bivariate distribution find the straight line for which the sum of the squares of the *normal* deviations is a minimum.

Let the straight line be $x \cos \alpha + y \sin \alpha = p$. Then we have to minimize the sum of squares $\sum f_i (x_i \cos \alpha + y_i \sin \alpha - p)^2$ by equating to zero its partial derivatives with respect to p and α. Show, from the first of the equations thus obtained, that the required line passes through the mean of the distribution. Then, taking the mean as origin, show from the second equation that α is given by

$$\tan 2\alpha = \frac{2\mu_{11}}{\sigma_x^2 - \sigma_y^2}.$$

Of the two directions at right angles, found from this equation, one makes the sum of squares a minimum, and the other a maximum, for lines through the mean of the distribution. (L. J. Reed, *Metron*, vol. I, 1921, part 3, pp. 54–61.)

7. The ranks of the same 15 students in Mathematics and Latin were as follows, the two numbers within brackets denoting the ranks of the same student: $(1, 10)$, $(2, 7)$, $(3, 2)$, $(4, 6)$, $(5, 4)$, $(6, 8)$, $(7, 3)$, $(8, 1)$, $(9, 11)$, $(10, 15)$, $(11, 9)$, $(12, 5)$, $(13, 14)$, $(14, 12)$, $(15, 13)$. Show that the rank correlation coefficient is 0.51.

8. The marks, x and y, gained by 1,000 students for theory and laboratory work respectively, are grouped with common class

x / y	42	47	52	57	62	67	72	77	82	Totals
52	3	9	19	4	—	—	—	—	—	35
57	9	26	37	25	6	—	—	—	—	103
62	10	38	74	45	19	6	—	—	—	192
67	4	20	59	96	54	23	7	—	—	263
72	—	4	30	54	74	43	9	—	—	214
77	—	—	7	18	31	50	19	5	—	130
82	—	—	—	2	5	13	15	8	3	46
87	—	—	—	—	—	2	5	8	2	17
Totals	26	97	226	244	189	137	55	21	5	1,000

interval of 5 marks for each variable, the frequencies for the various classes being shown in the correlation table above. The values of

x and y indicated are the mid-values of the classes. Show that the coefficient of correlation is 0·68, and the regression equation of y on x

$$y = 29\cdot7 + 0\cdot656x.$$

9. The logarithm of the m.g.f., $M(t_1, t_2)$, of § 31 is the cumulative function $K(t_1, t_2)$; and the cumulant κ_{rs} is the coefficient of $t_1^r t_2^s / r! \, s!$ in the expansion of $K(t_1, t_2)$ in powers of t_1 and t_2. Show that

$$\kappa_{11} = \mu_{11}, \quad \kappa_{21} = \mu_{21}, \quad \kappa_{31} = \mu_{31} - 3\mu_{20}\mu_{11}, \quad \kappa_{22} = \mu_{22} - \mu_{20}\mu_{02} - 2\mu_{11}^2.$$

10. By means of the identity

$$2 \sum fuv = \sum fu^2 + \sum fv^2 - \sum f(u - v)^2$$

and the fact that $u - v$ is constant along each of one set of the diagonal lines of a correlation table, the sum of products may be calculated without difficulty. Tabulate the quantities $u - v$, $(u - v)^2$, f and $f(u - v)^2$ for each diagonal line of the correlation table on p. 77, and deduce that $\sum f(u - v)^2 = 1147$. Using the values found for $\sum fu^2$ and $\sum fv^2$ verify that $\sum fuv = 1109$.

Deduce the same result from the identity

$$2 \sum fuv = \sum f(u + v)^2 - \sum fu^2 - \sum fv^2$$

using the other set of diagonal lines.

CHAPTER V

FURTHER CORRELATION THEORY.
CURVED REGRESSION LINES

33. Arrays. Linear regression

In the numerical illustration of § 29 all the values of x in any one vertical array were regarded as equal to the mid-value of x for that interval. Since the actual values of x vary over a range of 5 years, the results obtained cannot be regarded as more than a good approximation. A better approximation could be obtained by choosing a smaller class interval; but that would make the numerical work correspondingly heavier. For accurate work the class interval must be so small that all the values of x in a vertical array are either exactly or very nearly equal; and similarly all the y's in a horizontal array. The theoretical proof of certain formulae, however, is not made any more difficult by a choice of arrays sufficiently numerous to satisfy the above conditions; for our sums are just as easy to manage whether the number of arrays is large or small. And, if we are dealing with a continuous distribution, we may take infinitesimal class intervals, dx and dy, and replace our sums by definite integrals. We assume then that, in the ith vertical array, all the x's have the same value, x_i.

It will be convenient to extend our subscript notation as follows. Though the x's in the ith vertical array all have the same value, the y's are different. A typical pair of values in the array is (x_i, y_{ij}), represented by the point P_{ij}; and we shall denote its frequency by f_{ij}. Thus the first subscript, i, indicates the vertical array, while the second, j, indicates the position in that array. Denoting the total frequency in the ith array by n_i we have

$$n_i = \sum_j f_{ij}, \tag{1}$$

$\sum\limits_{j}$ indicating summation within that array. The mean value of y in this array will be denoted by $\bar{y}_i$. Thus

$$n_i \bar{y}_i = \sum_j f_{ij} y_{ij}. \tag{2}$$

The mean of the array is represented by the point G_i, whose co-ordinates are $(x_i, \bar{y}_i)$. H_i is the point with abscissa x_i on the line of regression of y on x. The mean, G, of the distribution also lies on this line of regression, and has coordinates $(\bar{x}, \bar{y})$ given by

$$N\bar{x} = \sum_i \sum_j f_{ij} x_i = \sum_i n_i x_i, \quad N\bar{y} = \sum_i \sum_j f_{ij} y_{ij} = \sum_i n_i \bar{y}_i. \quad (3)$$

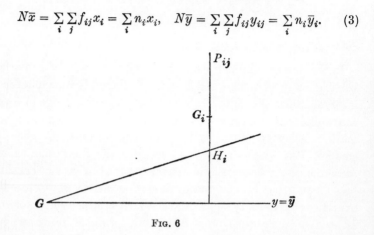

FIG. 6

The summation over the whole distribution is thus divided into two summations; for $\sum_j$ denotes summation for terms in the same vertical array, and then $\sum_i$ indicates summation of the result over all the arrays.

It is worth noting that the same equation is obtained for the line of regression of y on x, if each value of y in any array is replaced by the mean value of y in that array. For the equation of this regression line depends only on $\bar{x}$, $\bar{y}$, σ_x^2 and μ_{11}. And, since the above change does not affect the x's, $\bar{x}$ and σ_x^2 are unaltered by it. So also is $\bar{y}$, in virtue of the second equation (3). As for μ_{11} we see that

$$N\mu_{11} = \sum_i \sum_j f_{ij} x_i y_{ij} - N\overline{xy} = \sum_i n_i x_i \bar{y}_i - N\overline{xy}.$$

by (3), and is therefore unaltered by the above change of y's. Consequently, the regression line of y on x is unaltered. It follows that, if the means of the vertical arrays are collinear, their line must coincide with the line of regression of y on x. This regression is then said to be *linear*. Similarly, the regression of x on y is said to be linear if the means of the horizontal arrays all lie on the line of regression of x on y. It is possible for either regression to be linear, or both.

It was shown in §26 that $S_y^2 = \sigma_y^2(1-r^2)$ gives the mean square deviation of points from the line of regression of y on x. If this regression is linear the means of the vertical arrays lie on this line of regression; and, if further the vertical arrays all have the same variance, this common variance must be S_y^2. The S.D. of each array is then $\sigma_y\sqrt{(1-r^2)}$. When the vertical arrays all have the same variance, the regression of y on x is said to be *homoscedastic*.

34. Correlation ratios

Consider next the deviation $G_i P_{ij}$ of the point P_{ij} from the mean of the vertical array in which it lies. The sum of the squares of these deviations for the whole distribution is denoted by $N S_y'^2$, so that

$$N S_y'^2 = \sum_i \sum_j f_{ij}(y_{ij} - \bar{y}_i)^2. \tag{4}$$

Then, by analogy with §26 (22), the *correlation ratio*,* η_y, of y on x is defined by

$$\eta_y^2 = 1 - S_y'^2/\sigma_y^2, \tag{5}$$

η_y being regarded as positive. Thus

$$S_y'^2 = \sigma_y^2(1 - \eta_y^2), \tag{6}$$

which corresponds to §26 (21). From (5) it is clear that $\eta_y^2 \leqslant 1$. Also it is easy to show that $\eta_y^2 \geqslant r^2$. This is evident from a comparison of (5) with §26 (22), since $S_y'^2 \leqslant S_y^2$, the sum of the squares of the deviations in any array being least when they are measured from the mean of the array. Thus

$$1 \geqslant \eta_y^2 \geqslant r^2. \tag{7}$$

When the regression of y on x is linear, the straight line of means of arrays coincides with the line of regression, and η_y^2 is then equal to r^2. A non-zero value of $\eta_y^2 - r^2$ is thus associated with a departure of the regression from linearity.

* Some writers denote this 'ratio' by η_{yx}.

It is clear from (6) that the more nearly η_y^2 approaches unity, the smaller is $S_y'^2$, and therefore the closer are the points to the curve of means of the vertical arrays. If $\eta_y^2 = 1$, then $S_y'^2 = 0$, so that all the deviations are zero, and all the points lie on the curve of means. There is then a functional relation between x and y. We may therefore describe the correlation ratio η_y as *a measure of the degree to which the association between the variables approaches a functional relationship* expressible in the form $y = F(x)$, where $F(x)$ is a single-valued function of x. This should be compared with the corresponding interpretation of r in §26.

A convenient expression for η_y can be found, involving the s.d. σ_{my} of the means of the vertical arrays, each mean being weighted with the frequency n_i of that array. This s.d. is given by

$$N\sigma_{my}^2 = \sum_i n_i(\bar{y}_i - \bar{y})^2. \qquad (8)$$

Now

$$N\sigma_y^2 = \sum_i \sum_j f_{ij}(y_{ij} - \bar{y})^2$$
$$= \sum_i \sum_j f_{ij}[(y_{ij} - \bar{y}_i) + (\bar{y}_i - \bar{y})]^2$$
$$= \sum_i \sum_j f_{ij}(y_{ij} - \bar{y}_i)^2 + \sum_i n_i(\bar{y}_i - \bar{y})^2,$$

the sum of products being zero, since $\sum_j f_{ij}(y_{ij} - \bar{y}_i)$ vanishes for each array. The first sum on the right is $NS_y'^2$, in virtue of (4), and the second is $N\sigma_{my}^2$. Consequently the equation is equivalent to

$$\sigma_y^2 = S_y'^2 + \sigma_{my}^2. \qquad (9)$$

Thus the variance of the y's of the distribution is expressible as the sum of two parts, of which the first is the variance within the arrays, and the second the variance of the weighted means of the arrays. Also, comparing (9) with (6), we see that

$$\eta_y^2 = \sigma_{my}^2/\sigma_y^2,$$

and therefore

$$\eta_y = \sigma_{my}/\sigma_y. \qquad (10)$$

The correlation ratio η_y is therefore the ratio of the s.d. of the weighted means of the arrays of y's to the s.d. of all the y's of the distribution.

In the same way, by considering horizontal arrays in each of which the value of y is constant, we may define the correlation ratio η_x of x on y. For, if G_j is the mean of the jth horizontal array, with abscissa $\bar{x}_j$, and $NS_x'^2$ denotes the sum of the squares of the deviations of the points, each from the mean of the horizontal array in which it lies, η_x^2 is given by

$$S_x'^2 = \sigma_x^2(1 - \eta_x^2), \tag{11}$$

and, as before, we have the relations

$$1 \geqslant \eta_x^2 \geqslant r^2, \tag{12}$$

and
$$\eta_x = \sigma_{mx}/\sigma_x, \tag{13}$$

where σ_{mx} is the S.D. of the means of the horizontal arrays, each mean being weighted with the frequency n_j of that array.

35. Calculation of correlation ratios

Formula (10), giving the value of η_y^2, is clearly equivalent to

$$\eta_y^2 = \sum_i n_i(\bar{y}_i - \bar{y})^2 / N\sigma_y^2. \tag{14}$$

Now the sum in the numerator may be evaluated without calculating the deviations $\bar{y}_i - \bar{y}$. For

$$\sum_i n_i(\bar{y}_i - \bar{y})^2 = \sum_i n_i \bar{y}_i^2 - 2\bar{y}\sum_i n_i\bar{y}_i + N\bar{y}^2.$$

But
$$n_i\bar{y}_i = \sum_j f_{ij}y_{ij} = T_i,$$

where T_i is the sum of the y's in the ith vertical array. Hence

$$\sum_i n_i\bar{y}_i = \sum_i T_i = N\bar{y} = T, \tag{15}$$

where T is the sum of the y's for the whole distribution. We may therefore write

$$\sum_i n_i(\bar{y}_i - \bar{y})^2 = \sum_i \frac{T_i^2}{n_i} - 2\frac{T^2}{N} + \frac{T^2}{N} = \sum_i \frac{T_i^2}{n_i} - \frac{T^2}{N},$$

and (14) becomes

$$\eta_y^2 = \left(\sum_i \frac{T_i^2}{n_i} - \frac{T^2}{N}\right) \Big/ N\sigma_y^2. \tag{16}$$

Since the deviation from the mean is independent of the choice of origin, while numerator and denominator of (14) are altered in the same ratio by change of unit, the value of η_y^2 calculated from (16) is unaffected by change of origin and unit.

Example. Calculate the correlation ratios for the distribution of ages of bridegroom and bride given in § 29.

Working in terms of v and the 5-year unit we have

$$T = -371, \quad N = 1,000, \quad T^2/N = 137\cdot64.$$

The values of T_i are given in the row prefixed V. Hence

$$\sum_i \frac{T_i^2}{n_i} = \frac{(-28)^2}{17} + \frac{(-339)^2}{332} + \ldots + \frac{(26)^2}{7} = 876\cdot81,$$

$$N\sigma_v^2 = 1393\cdot4.$$

Consequently $\eta_y^2 = \dfrac{876\cdot81 - 137\cdot64}{1393\cdot4} = \dfrac{739\cdot17}{1393\cdot4} = 0\cdot5305,$

and $\eta_v = 0\cdot729$ nearly.

This is only slightly greater than the value $0\cdot726$ found for r.

The reader should verify, in the same way, that $\eta_x^2 = 0\cdot5504$, so that $\eta_x = 0\cdot742$. The departure of the regressions from linearity is only slight.

36. Other relations

We shall now prove that the mean square deviation of the weighted means of the vertical arrays from the line of regression of y on x is equal to $\sigma_y^2(\eta_y^2 - r^2)$, that is to say

$$\sum_i n_i(\bar{y}_i - Y_i)^2 = N\sigma_y^2(\eta_y^2 - r^2), \tag{17}$$

where Y_i is the estimate of y_i from the regression equation § 25 (16). For, subtraction of (6) from § 26 (21) shows that

$$N\sigma_y^2(\eta_y^2 - r^2) = N(S_y^2 - S_y'^2)$$
$$= \sum_i \sum_j f_{ij}[(y_{ij} - Y_i)^2 - (y_{ij} - \bar{y}_i)^2]$$
$$= \sum_i \sum_j f_{ij}[\{(y_{ij} - \bar{y}_i) + (\bar{y}_i - Y_i)\}^2 - (y_{ij} - \bar{y}_i)^2]$$
$$= \sum_i n_i(\bar{y}_i - Y_i)^2,$$

the sum of products vanishing, since $\sum_j f_{ij}(y_{ij} - \bar{y}_i)$ is zero for each array. Thus (17) has been established. The difference $\eta_y^2 - r^2$ is zero only when each of the deviations $\bar{y}_i - Y_i$ is zero. In that case the means of arrays all lie on the line of regression, and the regression of y on x is linear. We thus see again that a non-zero value of $\eta_y^2 - r^2$ is associated with departure of the regression from linearity.

We may prove, for use in a later chapter, that the sum of squares $\sum_i n_i(\bar{y}_i - \bar{y})^2$ may be resolved into two separate sums. Thus

$$\sum_i n_i(\bar{y}_i - \bar{y})^2 = \sum_i n_i[(\bar{y}_i - Y_i) + (Y_i - \bar{y})]^2$$
$$= \sum_i n_i(\bar{y}_i - Y_i)^2 + \sum_i n_i(Y_i - \bar{y})^2, \qquad (18)$$

the sum of products vanishing, since

$$\sum_i n_i(\bar{y}_i - Y_i)(Y_i - \bar{y}) = \sum_i \sum_j f_{ij}(y_{ij} - Y_i)(Y_i - \bar{y}) = 0$$

in virtue of § 27 (30). The sum in the first member of (18) is $N\sigma_{my}^2$, or $N\eta_y^2\sigma_y^2$ by (10). The first sum on the right of (18) has just been proved equal to $N\sigma_y^2(\eta_y^2 - r^2)$; and the final sum has the value $Nr^2\sigma_y^2$, in virtue of § 27 (32). The equation (18) thus corresponds to the identity

$$N\eta_y^2\sigma_y^2 = N\sigma_y^2(\eta_y^2 - r^2) + Nr^2\sigma_y^2. \qquad (19)$$

37. Continuous distributions

The preceding proofs may be adapted to a continuous distribution by an appropriate change of notation, and a replacement of sums by definite integrals. The vertical array, whose abscissa is x, we assume to be of infinitesimal breadth dx. Then, if $f(x, y)$ is the relative frequency density, the relative frequency for this array is

$$\int [f(x, y) dx] dy = f_1(x) dx, \qquad (20)$$

where
$$f_1(x) = \int f(x, y) dy, \qquad (21)$$

the integration with respect to y extending over the whole of that array. The mean $\bar{y}_x$ of the y's in the array is given by

$$\bar{y}_x f_1(x) = \int y f(x, y) dy, \qquad (22)$$

and this is the equation of the curve of means of the vertical arrays. The mean of the whole distribution has coordinates

$$\left.\begin{array}{l} \bar{x} = \iint xf(x,y)\,dx\,dy = \int xf_1(x)\,dx, \\[2mm] \bar{y} = \iint yf(x,y)\,dx\,dy = \int \bar{y}_x f_1(x)\,dx, \end{array}\right\} \tag{23}$$

integration with respect to x including all the vertical arrays. The above relations correspond to the equations (1), (2) and (3) for a discrete distribution.

The mean square deviation of the values of y, from the means of the arrays in which they lie, is given by

$$S_y'^2 = \iint (y - \bar{y}_x)^2 f(x,y)\,dx\,dy, \tag{24}$$

and η_y is defined as before by an equation of the form (5) or (6). The variance of the weighted means of the vertical arrays is

$$\sigma_{my}^2 = \int (\bar{y}_x - \bar{y})^2 f_1(x)\,dx, \tag{25}$$

and, as before, this may be expressed in terms of σ_y^2 and $S_y'^2$, since

$$\begin{aligned} \sigma_y^2 &= \iint (y - \bar{y})^2 f(x,y)\,dx\,dy \\ &= \iint [(y - \bar{y}_x) + (\bar{y}_x - \bar{y})]^2 f(x,y)\,dx\,dy \\ &= S_y'^2 + \sigma_{my}^2, \end{aligned} \tag{26}$$

the integral of the product being equal to zero; and from this it follows, as in the case of a discrete distribution, that

$$\eta_y = \sigma_{my}/\sigma_y.$$

Formulae corresponding to those of § 36 may be similarly established. For, by subtraction of (24) from the result in the Example of § 26,

$$\begin{aligned} \sigma_y^2(\eta_y^2 - r^2) &= S_y^2 - S_y'^2 = \iint [(y - Y)^2 - (y - \bar{y}_x)^2] f(x,y)\,dx\,dy \\ &= \iint [\{(y - \bar{y}_x) + (\bar{y}_x - Y)\}^2 - (y - \bar{y}_x)^2] f(x,y)\,dx\,dy \\ &= \iint (\bar{y}_x - Y)^2 f(x,y)\,dx\,dy = \int (\bar{y}_x - Y)^2 f_1(x)\,dx, \end{aligned}$$

and this is the mean square deviation of the weighted means of arrays from the line of regression of y on x. And, corresponding to (18), we have the resolution

$$\int (\bar{y}_x - \bar{y})^2 f_1(x)\, dx = \int [(\bar{y}_x - Y) + (Y - \bar{y})]^2 f_1(x)\, dx$$
$$= \int (\bar{y}_x - Y)^2 f_1(x)\, dx + \int (Y - \bar{y})^2 f_1(x)\, dx,$$

the various integrals corresponding to terms of the identity

$$\eta_y^2 \sigma_y^2 = \sigma_y^2 (\eta_y^2 - r^2) + r^2 \sigma_y^2.$$

38. Bivariate normal distribution

We shall now consider briefly the continuous bivariate distribution, which is a generalization of the normal distribution discussed in Chapter III. It may be introduced simply as follows.* Assume first that the variable x is normally distributed with s.d. σ_1. Then, if the variable is measured from its mean, the probability that a random value of x will fall in the interval dx is

$$dP_1 = \frac{dx}{\sigma_1 \sqrt{(2\pi)}} \exp\left(-\frac{x^2}{2\sigma_1^2}\right).$$

Assume next that the regression of y on x is linear and homoscedastic. Then, if σ_2 is the s.d. of y in the distribution, the common variance of the arrays of y's is $\sigma_2^2(1 - \rho^2)$, where ρ is the† coefficient of correlation between the variables. Finally, assume that each array of y's is normally distributed. Then, since the mean of each array is on the line of regression

$$y = \rho x \sigma_2 / \sigma_1, \tag{27}$$

and the variance of each array is as stated, the probability that a value of y, taken at random in an assigned vertical array, will fall in the interval dy is

$$dP_2 = \frac{dy}{\sigma_2 \sqrt{\{2\pi(1 - \rho^2)\}}} \exp\left[-\frac{1}{2\sigma_2^2(1 - \rho^2)}\left(y - \frac{\rho x \sigma_2}{\sigma_1}\right)^2\right].$$

* Cf. Rietz, 1927, 2, pp. 104–7.

† In the theory of sampling, r refers to the sample, and σ_1, σ_2, ρ to the population from which the sample is drawn.

By the theorem of compound probability, the chance of a pair of values (x, y) falling in the elementary rectangle $dx\,dy$ is

$$dP_1\,dP_2 = \frac{dx\,dy}{2\pi\sigma_1\sigma_2\sqrt{(1-\rho^2)}} \exp\left[-\frac{1}{2(1-\rho^2)}\left\{\frac{x^2}{\sigma_1^2} - \frac{2\rho xy}{\sigma_1\sigma_2} + \frac{y^2}{\sigma_2^2}\right\}\right].$$

The probability density $\phi(x, y)$ for the distribution is therefore

$$\phi(x, y) = \frac{1}{2\pi\sigma_1\sigma_2\sqrt{(1-\rho^2)}} \exp\left[-\frac{1}{2(1-\rho^2)}\left\{\frac{x^2}{\sigma_1^2} - \frac{2\rho xy}{\sigma_1\sigma_2} + \frac{y^2}{\sigma_2^2}\right\}\right]. \quad (28)$$

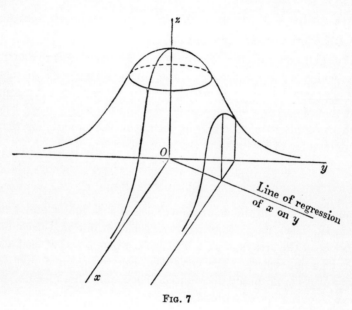

FIG. 7

Such a distribution is called a *bivariate normal* distribution, and the variables are said to be *normally correlated*. The surface $z = \phi(x, y)$ is the *normal correlation surface*. Since (28) is of the same form in x as in y, we may conclude that the regression of x on y is also linear, the variance of each array of x's being $\sigma_1^2(1 - \rho^2)$. The values of x in each such array are normally distributed, with mean on the line of regression of x on y, whose equation is

$$x = \rho y \sigma_1/\sigma_2. \quad (29)$$

The m.g.f. for the distribution is found below,[*] and from it the first few moments are calculated.

The nature of the normal correlation surface is indicated in the diagram. The curves along which the probability density (or the relative frequency density) is constant are the homothetic ellipses

$$\frac{x^2}{\sigma_1^2} - \frac{2\rho xy}{\sigma_1 \sigma_2} + \frac{y^2}{\sigma_2^2} = \lambda^2. \tag{30}$$

With respect to these the line of regression of y on x is conjugate to the y-axis. For the locus of the mid-points of chords $x =$ const. is the straight line (27). Similarly the line of regression of x on y is conjugate to the x-axis.[†]

39. Intraclass correlation

Let us now consider the correlation between the measures of some common characteristic for pairs of members of the same family or class. For example, we may be interested in the correlation between the weights of brothers, or the heights of sisters. The relation between two members of the same family is a reciprocal one; for, if P belongs to the same family as Q, then Q belongs to the same family as P. Each pair of members, P and Q, will therefore contribute two entries to the correlation table. In one of them x will be the measure of the characteristic for P, and y the measure for Q; in the other, x will be the measure for Q, and y for P. The table will thus be symmetrical.

We shall consider only the case in which each of the h families has the same number, k, of members. Then there are $k(k-1)$ pairs of values for each family, and

$$N = hk(k-1) \tag{31}$$

gives the total number of pairs of values in the table. Let x_{ij} denote the measure of the characteristic for the jth member of the ith family. Thus the first subscript indicates the family, and the second the particular member of that family. Consequently i takes the values $1, 2, ..., h$, and j the values $1, 2, ..., k$. The x's and the y's

* See Ex. V, 3.
† For further properties of this distribution see Exx. 4, 5 and 6 at the end of this chapter.

of the correlation table are the hk values x_{ij} in different orders. Any one value x_{ij}, occurring as an x in the table, will have as its y each of the other $k-1$ values for the same family. Thus each value x_{ij} occurs $k-1$ times as an x; and the mean $\bar{x}$ for the bivariate distribution is therefore

$$\bar{x} = \frac{(k-1)\sum_i\sum_j x_{ij}}{hk(k-1)} = \frac{1}{hk}\sum_i\sum_j x_{ij}, \tag{32}$$

and, since the table is symmetrical, $\bar{y}$ has the same value. The variance σ_x^2 of the x's is the same as the variance of the hk values x_{ij}, since each of these occurs the same number of times in the distribution. Thus

$$\sigma_x^2 = \frac{1}{hk}\sum_i\sum_j (x_{ij}-\bar{x})^2, \tag{33}$$

and σ_y^2 has the same value, in virtue of symmetry. We may denote this common variance by σ^2, so that

$$\sigma^2 = \sigma_x^2 = \sigma_y^2. \tag{34}$$

The *coefficient of intraclass correlation*, r, is given by the usual formula, which in this case is equivalent to

$$hk(k-1)\sigma^2 r = \sum_i\sum_j\sum_l (x_{ij}-\bar{x})(x_{il}-\bar{x}) \tag{35}$$

$$(j,l = 1, 2, ..., k; \; j \neq l; \; i = 1, 2, ..., h).$$

The sum is a triple one; for the product $(x_{ij}-\bar{x})(x_{il}-\bar{x})$ is the product of the deviations from the mean for the jth and lth members of the ith family, and the sum must include all such products for different values of j and l, and for all the families. We carry out first the summation with respect to l, observing that l takes all integral values from 1 to k except the value j. Thus the sum of the terms x_{il} includes all values for the ith family except x_{ij}, and may therefore be written $k\bar{x}_i - x_{ij}$, where $\bar{x}_i$ is the mean for the ith family. The sum of the terms $\bar{x}$ for the $k-1$ values of l is $(k-1)\bar{x}$. The triple sum in (35) may therefore be written as the double sum

$$\sum_i\sum_j (x_{ij}-\bar{x})[k\bar{x}_i - x_{ij} - (k-1)\bar{x}]$$
$$= k\sum_i\sum_j (x_{ij}-\bar{x})(\bar{x}_i-\bar{x}) - \sum_i\sum_j (x_{ij}-\bar{x})^2.$$

Now by (33) the last sum on the right is equal to $hk\sigma^2$. To evaluate the other we first carry out the summation with respect to j. Thus

$$k \sum_i \sum_j (x_{ij} - \bar{x})(\bar{x}_i - \bar{x}) = k \sum_i (\bar{x}_i - \bar{x})(k\bar{x}_i - k\bar{x})$$
$$= k^2 \sum_i (\bar{x}_i - \bar{x})^2 = k^2 h \sigma_m^2,$$

where σ_m^2 is the variance of the means of the families. Substituting these values in (35) we obtain

$$hk(k-1)\sigma^2 r = hk^2 \sigma_m^2 - hk\sigma^2.$$

The coefficient of intraclass correlation is therefore given by the formula

$$r = \frac{k\sigma_m^2 - \sigma^2}{(k-1)\sigma^2}. \tag{36}$$

Curved Regression Lines

40. Polynomial regression. Normal equations

We have seen that the line of regression of y on x gives the best representation of the behaviour of y with change of x, that can be given by a straight line, the term 'best' indicating that the sum of the squares of the deviations (or 'residuals') is less for this straight line than for any other. But it is often apparent from the data that the regression of y on x is far from linear; and it may be desirable to find an equation of regression which affords a better representation of the behaviour of y than can be given by a straight line. The simplest type of non-linear regression equation is that in which one of the variables is expressed as a polynomial in the other. Polynomial regression of y on x is therefore represented by an equation of the form

$$y = b_0 + b_1 x + b_2 x^2 + \ldots + b_k x^k, \tag{37}$$

in which the coefficients b_s are constants. Our problem is to determine these constants so that the sum of the squares of the residuals is a minimum. The choice of the degree, k, of the polynomial is at our disposal. If there are n different pairs of values in the distribution, we can make the curve pass through all the representative points by choosing $k = n - 1$. But, to keep the arithmetical work reasonably simple, k must be fairly small, say 2, 3, or 4. The distribution of points in the scatter diagram will frequently suggest the shape of the curve of regression, and thus a suitable value for k. Having decided on the value of k, we determine the coefficients b_s by the method of least squares.

With the notation of § 23 suppose there are n different pairs of values, f_i being the frequency of the pair (x_i, y_i), and the total frequency N being given by

$$N = \sum_{i=1}^{n} f_i. \tag{38}$$

Let H_i be the point on the curve (37) with abscissa x_i. Then its ordinate Y_i has the value

$$Y_i = b_0 + b_1 x_i + \ldots + b_k x_i^k = \sum_{s=0}^{k} b_s x_i^s, \tag{39}$$

and the deviation of the point P_i from the curve of regression is

$$H_i P_i = y_i - Y_i. \tag{40}$$

The sum of the squares of the deviations is

$$N S^2 = \sum_i f_i (y_i - Y_i)^2. \tag{41}$$

We have to choose the coefficients b_s so that this sum is a minimum; and this is done by equating to zero the partial derivatives of S^2 with respect to these coefficients. We thus obtain the $k + 1$ *normal equations*

$$\sum_i f_i x_i^s (y_i - Y_i) = 0 \tag{42}$$

$$(i = 1, 2, \ldots, n; \ s = 0, 1, 2, \ldots, k).$$

Written separately, and in terms of the values of the given distribution, these are

$$\left.\begin{aligned}
\sum_i f_i (y_i - b_0 - b_1 x_i - \ldots - b_k x_i^k) &= 0, \\
\sum_i f_i x_i (y_i - b_0 - b_1 x_i - \ldots - b_k x_i^k) &= 0, \\
\cdots\cdots\cdots\cdots\cdots\cdots\cdots\cdots\cdots\cdots\cdots\cdots\cdots\cdots, \\
\sum_i f_i x_i^k (y_i - b_0 - b_1 x_i - \ldots - b_k x_i^k) &= 0.
\end{aligned}\right\} \tag{42'}$$

The coefficients b_s are determined from these $k+1$ equations, which involve the sums of powers of x_i from $\sum x_i$ to $\sum_i x_i^{2k}$, and sums of products from $\sum y_i x_i^0$ to $\sum y_i x_i^k$. For the particular case $k = 1$ these equations are the same as § 25 (12).

When the values x_i correspond to equal increments, h, and the distribution of the frequencies f_i is symmetrical about $\bar{x}$, the equations (42') may be much simplified. Suppose first that n is odd and equal to $2m+1$. Then, by taking the origin of x at the middle value of that variable, and the common increment h as unit of measurement, we have for the values of x in the distribution

$$-m, \ -(m-1), \ ..., \ -1, \ 0, \ 1, \ ..., \ (m-1), \ m.$$

Hence, owing to the symmetry of the distribution of f's, the sums of the odd powers of x are all zero. The sums of the even powers may be written down from tables or algebraical formulae. If, however, n is even and equal to $2m$, we take the origin of x at the mean of the middle pair of values, and $\frac{1}{2}h$ as the new unit. The values of x then become

$$-(2m-1), \ ..., \ -3, \ -1, \ 1, \ 3, \ ..., \ (2m-1),$$

and the sums of the odd powers of x vanish as before. Moreover, the equations (42') then consist of two groups, one involving $b_0, b_2, ...$ and the other $b_1, b_3, ...$ Their solution is thus simplified.

Example. Fit a parabolic curve of regression of y on x to the seven pairs of values

$$x: \quad 1\cdot0 \quad 1\cdot5 \quad 2\cdot0 \quad 2\cdot5 \quad 3\cdot0 \quad 3\cdot5 \quad 4\cdot0$$
$$y: \quad 1\cdot1 \quad 1\cdot3 \quad 1\cdot6 \quad 2\cdot0 \quad 2\cdot7 \quad 3\cdot4 \quad 4\cdot1$$

The dot diagram of the seven pairs of values suggests a parabola. Since n is odd, and the values of x correspond to equal increments $0\cdot5$, we take the origin at the middle value $2\cdot5$ with $0\cdot5$ as the new unit. This is equivalent to the transformation

$$u = 2x - 5.$$

Each frequency f_i is unity. The sums of odd powers of u are zero, and the calculation may be arranged as in the accompanying table. The normal equations are

$$16\cdot2 = 7b_0 + 28b_2, \quad 14\cdot3 = 28b_1, \quad 69\cdot9 = 28b_0 + 196b_2,$$

which give immediately

$$b_0 = 2\cdot07, \quad b_1 = 0\cdot511, \quad b_2 = 0\cdot061.$$

The regression equation is therefore

$$y = 2 \cdot 07 + 0 \cdot 511u + 0 \cdot 061u^2$$
$$= 2 \cdot 07 + 0 \cdot 511(2x - 5) + 0 \cdot 061(2x - 5)^2,$$

which simplifies to $\qquad y = 1 \cdot 04 - 0 \cdot 20x + 0 \cdot 24x^2.$

x	u	y	u^2	u^4	uy	u^2y
1·0	−3	1·1	9	81	−3·3	9·9
1·5	−2	1·3	4	16	−2·6	5·2
2·0	−1	1·6	1	1	−1·6	1·6
2·5	0	2·0	0	0	—	—
3·0	1	2·7	1	1	2·7	2·7
3·5	2	3·4	4	16	6·8	13·6
4·0	3	4·1	9	81	12·3	36·9
Totals	—	16·2	28	196	14·3	69·9

41. Index of correlation

When considering the line of regression of y on x, we saw that $|r|$ is the coefficient of correlation between the values y_i and their estimates Y_i found from the regression equation, and that this coefficient is connected with the mean square deviation from the line of regression by the formula $S_y^2 = \sigma_y^2(1 - r^2)$. We propose to show that a corresponding relation holds for polynomial regression. Multiplying the normal equations $(42')$ by $b_0, b_1, \ldots, b_k$ respectively and adding we obtain, in virtue of (39),

$$\sum_i f_i Y_i (y_i - Y_i) = 0. \tag{43}$$

Consequently (41) is equivalent to

$$NS^2 = \sum_i f_i y_i (y_i - Y_i) = \sum_i f_i y_i^2 - \sum_i f_i Y_i^2. \tag{44}$$

From the first of the normal equations, $\sum_i f_i(y_i - Y_i) = 0$, it follows that the y's and the Y's have the same mean. If this is taken as origin for both variables, their variances are given by

$$N\sigma_y^2 = \sum_i f_i y_i^2, \quad N\sigma_Y^2 = \sum_i f_i Y_i^2,$$

and the coefficient of correlation, R, between the y_i and the Y_i by

$$R\sigma_y \sigma_Y = \frac{1}{N} \sum_i f_i y_i Y_i = \frac{1}{N} \sum_i f_i Y_i^2 = \sigma_Y^2$$

in virtue of (43), so that

$$R\sigma_y = \sigma_Y. \tag{45}$$

Consequently (44) is equivalent to

$$S^2 = \sigma_y^2(1 - R^2), \tag{46}$$

which is analogous to § 26 (21).

The coefficient R is thus an indication of the closeness with which the points P_i of the scatter diagram approximate to the regression curve (37). If $R = 1$, then $S^2 = 0$, and all the points P_i lie on the curve. For this reason R is often referred to as the *index of correlation* for the regression curve (37). With each curve of regression there is associated such an index, given by

$$R^2 = 1 - S^2/\sigma_y^2, \tag{47}$$

where S^2 is the mean square deviation of the points P_i from the curve of regression. In virtue of (43) and (44) NS^2 in the case of polynomial regression is given by

$$NS^2 = \sum_i f_i y_i^2 - \sum_{s=0}^{k} \sum_{i=1}^{n} f_i b_s y_i x_i^s. \tag{48}$$

Further, the relations (30) and (31) of § 27 hold also for polynomial regression. For, in virtue of (43) and the first of the normal equations, we have

$$\sum_i f_i (y_i - Y_i)(Y_i - \bar{y}) = 0. \tag{49}$$

Then the sum of the squares of the deviations of the y's from their mean may be expressed

$$\sum_i f_i (y_i - \bar{y})^2 = \sum f_i [(y_i - Y_i) + (Y_i - \bar{y})]^2$$
$$= \sum f_i (y_i - Y_i)^2 + \sum f_i (Y_i - \bar{y})^2, \tag{50}$$

in consequence of (49). This equation corresponds to the identity

$$N\sigma_y^2 = N\sigma_y^2(1 - R^2) + NR^2\sigma_y^2,$$

as is evident from (45) and (46).

Example. Prove, as in § 36, that $\sigma_y^2(\eta_y^2 - R^2)$ is the mean square deviation of the weighted means of the vertical arrays from the curve of regression. Also that, if n_i is the frequency in the ith vertical array,

$$\sum n_i (\bar{y}_i - \bar{y})^2 = \sum n_i (\bar{y}_i - Y_i)^2 + \sum n_i (Y_i - \bar{y})^2,$$

corresponding to the identity

$$N\eta_y^2 \sigma_y^2 = N\sigma_y^2 (\eta_y^2 - R^2) + NR^2 \sigma_y^2.$$

42. Some related regressions

A plotting of the values of x against the logarithms of the corresponding values of y may indicate an approximation to a linear relation between x and $\log y$. In such a case we find a regression equation of the form

$$y = ca^x, \tag{51}$$

where c and a are constants. For this relation is equivalent to

$$\log y = \log c + x \log a.$$

If then we find the line of regression of $\log y$ on x, the constants of the equation determine both c and a.

Similarly, if the plotting of $\log x$ against $\log y$ indicates that the relation between these quantities approximates to linear, we find a regression equation of the form

$$y = cx^b, \tag{52}$$

which is equivalent to

$$\log y = \log c + b \log x.$$

We have therefore to find the line of regression of $\log y$ on $\log x$, and the constants of the equation determine c and b.

We might also find a regression equation of the form

$$y = ca^{f(x)}, \tag{53}$$

where $f(x)$ is a polynomial in x. For this is equivalent to

$$\log y = \log c + (\log a) f(x),$$

and therefore requires a polynomial regression of $\log y$ on x.

More generally, the principle of least squares may be employed to find a regression equation of the form*

$$y = b_1 X_1 + b_2 X_2 + \dots + b_p X_p,$$

where $X_1, \dots, X_p$ are any functions of the independent variable x. The argument used in § 40 leads to the normal equations

$$\sum_i f_i X_1 (y_i - Y_i) = 0, \quad \dots, \quad \sum_i f_i X_p (y_i - Y_i) = 0,$$

---

* Cf. Fisher, *Annals of Eugenics*, vol. IX, part 3, p. 238 (1939).

for the determination of the coefficients b_s. Multiplying these equations by $b_1, \ldots, b_p$ respectively and adding, we find

$$\sum_i f_i Y_i (y_i - Y_i) = 0,$$

so that the sum of squares of the deviations from the regression curve is

$$NS^2 = \sum f_i (y_i - Y_i)^2 = \sum f_i y_i (y_i - Y_i)$$
$$= \sum f_i y_i^2 - \sum f_i Y_i^2,$$

as in § 41. If one of the quantities X_s is taken as unity (or a constant), the mean of the y's is equal to that of the Y's. The argument of § 41 then applies to the present case, leading to (45), (46), (49) and (50).

COLLATERAL READING

YULE and KENDALL, 1937, 1, chapters XII, XIII and XV–XVIII.

KENNEY, 1939, 3, part I, chapter VII; part II, chapter IV.

EZEKIEL, 1930, 2, chapters VI and VII.

AITKEN, 1939, 1, chapters V and VI.

RIDER, 1939, 6, §§ 20, 21 and 25–28.

RIETZ (ed.), 1924, 1, pp. 129–38.

JONES, 1924, 3, chapter XIX.

CAMP, 1934, 1, part I, chapter X.

MILLS, 1938, 1, chapter XII.

PLUMMER, 1940, 1, chapter V.

RIETZ, 1927, 2, §§ 32, 37 and 38.

GOULDEN, 1939, 2, chapter XIV.

COOLIDGE, 1925, 2, chapters VIII and IX.

TIPPETT, 1931, 2, chapter IX.

HARRIS, 1913, 3.

EXAMPLES V

1. The heights of brothers, in five families of three each, are as follows: (67, 68, 69), (68 ,68, 71), (68, 70, 72), (70, 70, 73) and (71, 72, 73) inches. Show that the mean heights for the families are 68, 69, 70, 71, 72 inches, and the general mean $\bar{x} = 70$. Also $\sigma^2 = 18/5$, $\sigma_m^2 = 2$, and the coefficient of intraclass correlation of heights for the five families is $1/3$.

2. Verify that, for the distribution of Ex. IV, 8, the correlation ratios are $\eta_x = 0{\cdot}695$ and $\eta_y = 0{\cdot}685$ approximately.

3. *Moments and m.g.f. for the bivariate normal distribution.* The moments may be deduced from the m.g.f. as defined in § 31. To calculate this function let σ_1 and σ_2 be the s.d.'s of x and y in the distribution, and ρ the correlation between these variates. Then the m.g.f. is given by

$$M(t_1, t_2) = c \int_{-\infty}^{\infty} \int_{-\infty}^{\infty} \exp\left[t_1 x + t_2 y - \frac{1}{2(1-\rho^2)} \left(\frac{x^2}{\sigma_1^2} - \frac{2\rho xy}{\sigma_1 \sigma_2} + \frac{y^2}{\sigma_2^2} \right) \right] dx\, dy,$$

where
$$c = \frac{1}{2\pi \sigma_1 \sigma_2 \sqrt{(1-\rho^2)}}.$$

By transforming the exponential we may write this:

$$M(t_1, t_2) = c \int_{-\infty}^{\infty} \int_{-\infty}^{\infty} \exp\left[t_2 y - \frac{1}{2\sigma_2^2(1-\rho^2)} \{ y^2 - (\rho y + \overline{1-\rho^2}\, t_1 \sigma_1 \sigma_2)^2 \} \right.$$
$$\left. - \frac{1}{2\sigma_1^2(1-\rho^2)} \left\{ x - \frac{\rho \sigma_1 y}{\sigma_2} - (1-\rho^2) t_1 \sigma_1^2 \right\}^2 \right] dy\, dx.$$

Carrying out the integration with respect to x, and rearranging the exponential, we have

$$M(t_1, t_2) = \frac{1}{\sigma_2 \sqrt{(2\pi)}} \exp\left[\tfrac{1}{2}(\sigma_1^2 t_1^2 + 2\rho \sigma_1 \sigma_2 t_1 t_2 + \sigma_2^2 t_2^2) \right]$$
$$\times \int_{-\infty}^{\infty} \exp\left[-\frac{1}{2\sigma_2^2}(y - \rho \sigma_1 \sigma_2 t_1 - t_2 \sigma_2^2)^2 \right] dy$$
$$= \exp\left[\tfrac{1}{2}(\sigma_1^2 t_1^2 + 2\rho \sigma_1 \sigma_2 t_1 t_2 + \sigma_2^2 t_2^2) \right],$$

which is the required m.g.f.

To calculate the various moments we have only to expand this function in powers of t_1 and t_2. The expansion is

$$1 + \tfrac{1}{2}(\sigma_1^2 t_1^2 + 2\rho \sigma_1 \sigma_2 t_1 t_2 + \sigma_2^2 t_2^2) + \tfrac{1}{4}(\sigma_1^2 t_1^2 + 2\rho \sigma_1 \sigma_2 t_1 t_2 + \sigma_2^2 t_2^2)^2/2! + \dots.$$

Since μ_{rs} is the coefficient of $t_1^r t_2^s/r!\,s!$ in this expansion we have

$$\mu_{20} = \sigma_1^2, \quad \mu_{11} = \rho \sigma_1 \sigma_2, \quad \mu_{02} = \sigma_2^2, \quad \mu_{31} = 3\rho \sigma_1^3 \sigma_2,$$
$$\mu_{22} = (1 + 2\rho^2)\sigma_1^2 \sigma_2^2, \quad \mu_{13} = 3\rho \sigma_1 \sigma_2^3, \quad \mu_{40} = 3\sigma_1^4, \quad \mu_{04} = 3\sigma_2^4,$$

and so on.

4. Show that the area of the ellipse, § 38 (30), of constant probability density is $\pi \lambda^2 \sigma_1 \sigma_2/\sqrt{(1-\rho^2)}$; and hence that the area of the strip between the ellipses corresponding to the parameter values λ and $\lambda + d\lambda$ is $2\pi\lambda\, d\lambda\, \sigma_1\sigma_2/\sqrt{(1-\rho^2)}$. Deduce the probability

$$\exp\left[-\lambda^2/2(1-\rho^2)\right]\lambda\, d\lambda/(1-\rho^2)$$

that a pair of values (x, y) chosen at random, will be represented by a point inside this strip; and hence by integration the probability that the point will fall inside the ellipse λ is

$$\frac{1}{1-\rho^2}\int_0^\lambda \lambda\exp\left[-\frac{\lambda^2}{2(1-\rho^2)}\right]d\lambda = 1-\exp\left[-\frac{\lambda^2}{2(1-\rho^2)}\right].$$

If this probability is $\frac{1}{2}$, then $\lambda^2 = 1\cdot 3863(1-\rho^2)$.

Also show that, for a given value of $d\lambda$, the probability that the point (x, y) will fall in the strip between the ellipses λ and $\lambda + d\lambda$ is a maximum when $\lambda^2 = 1-\rho^2$. This determines the 'ellipse of maximum probability'. (Cf. Rietz, 1927, 2, pp. 108–10.)

5. The variates x and y are normally correlated, and ξ, η are defined by

$$\xi = x\cos\theta + y\sin\theta, \quad \eta = y\cos\theta - x\sin\theta.$$

Show that ξ, η will be uncorrelated if

$$\tan 2\theta = 2\rho\sigma_1\sigma_2/(\sigma_1^2 - \sigma_2^2).$$

The above transformation corresponds to a rotation of rectangular axes of coordinates; and θ determines the directions of the principal axes of the ellipses of constant density.

Show that, if ξ, η are thus uncorrelated, and σ_ξ^2, σ_η^2 are their variances, then

$$\sigma_\xi\sigma_\eta = \sigma_1\sigma_2\sqrt{(1-\rho^2)}, \quad \sigma_\xi^2 + \sigma_\eta^2 = \sigma_1^2 + \sigma_2^2.$$

6. Show that, if ξ, η are independent normal variates, and x, y are defined by

$$x = \xi\cos\theta + \eta\sin\theta, \quad y = \eta\cos\theta - \xi\sin\theta,$$

the coefficient of correlation between x and y is given by (cf. Ex. 5)

$$r^2 = 1 - \frac{4\sigma_\xi^2\sigma_\eta^2}{(\sigma_\xi^2 - \sigma_\eta^2)^2\sin^2 2\theta + 4\sigma_\xi^2\sigma_\eta^2},$$

which is numerically greatest when $\theta = \pm\frac{1}{4}\pi$. The extreme values of r are $\pm(\sigma_\xi^2 - \sigma_\eta^2)/(\sigma_\xi^2 + \sigma_\eta^2)$.

Also show that the points of inflexion of sections of the normal correlation surface, by planes through the z-axis, lie on the elliptic cylinder $\xi^2\sigma_\eta^2 + \eta^2\sigma_\xi^2 = \sigma_\xi^2\sigma_\eta^2$. (Cf. Yule, 1897, 1.)

7. The profits, £y, of a certain company in the xth year of its life are given by

$$x:\quad 1\quad\quad 2\quad\quad 3\quad\quad 4\quad\quad 5$$
$$y:\quad 1250\quad 1400\quad 1650\quad 1950\quad 2300$$

Taking $u = x - 3$ and $v = (y - 1650)/50$, show that the parabolic regression of v on u is

$$v + 0 \cdot 086 = 5 \cdot 30u + 0 \cdot 643u^2,$$

and deduce that the parabolic regression of y on x is

$$y = 1140 + 72 \cdot 14x + 32 \cdot 14x^2.$$

8. Let x, y be normally correlated variates with zero means as in §38. Writing

$$w = \frac{x}{\sigma_1}, \quad z = \frac{1}{\sqrt{(1-\rho^2)}}\left(\frac{y}{\sigma_2} - \frac{\rho x}{\sigma_1}\right),$$

show that

$$\frac{\partial(w, z)}{\partial(x, y)} = \frac{1}{\sigma_1\sigma_2\sqrt{(1-\rho^2)}}$$

and

$$w^2 + z^2 = \frac{1}{1-\rho^2}\left(\frac{x^2}{\sigma_1^2} - \frac{2\rho xy}{\sigma_1\sigma_2} + \frac{y^2}{\sigma_2^2}\right).$$

Deduce that the joint probability differential of w and z is

$$dP = \frac{1}{2\pi}\exp\left[-\tfrac{1}{2}(w^2 + z^2)\right]dw\,dz,$$

and hence that w, z are independent normal variates, with zero means and unit S.D.'s. In other words w and z are independent standard normal variates.

CHAPTER VI

THEORY OF SIMPLE SAMPLING

43. Random sampling from a population

In order to examine a large population with respect to a specified characteristic, the statistician chooses a sample of individuals from that population and, from the properties of the sample relating to the given characteristic, he endeavours to estimate those of the population. Suppose that the characteristic considered is the height of the individual. Then the assemblage of heights of all the individuals in the population is called a *population* or *universe* of heights, and those of the individuals in the sample is a *sample* of heights from that population. Similarly, we might consider populations of weights, wages, yields of grain, etc. In the same way, if our consideration is the percentage of male births in a very large population of births, we may be obliged, in estimating this percentage, to confine our attention to the data provided by a sample of such births. The theory of sampling is concerned, first, with estimating the properties of the population from those of the sample, and secondly, with gauging the precision of the estimates, i.e. with ascertaining the deviations from the true values that may be expected in the estimates obtained.

Fundamental to the theory is the concept of *random sampling*. This is defined by the property that, in the selection of an individual from the population, each member of the population has the same chance of being chosen.* Statisticians have developed techniques for ensuring, as far as possible, that their sampling is random; but the reader who wishes to study the details of these techniques must consult other works.†

SAMPLING OF ATTRIBUTES

44. Simple sampling of attributes

In the sampling of attributes, as distinct from the sampling of values of a variable such as height, we are concerned only with the

---

* See also Kendall, 1941, 2.

† E.g. Yule and Kendall, 1937, 1, pp. 336–46.

possession or non-possession of some specified attribute or character-istic by the individual selected in sampling. For instance, in sampling from births we may be concerned only whether the baby is male or not. In sampling from a population of men our consideration may be whether they are smokers or non-smokers. The choosing of an individual in sampling may be called a 'trial', and the possession of the specified attribute by the individual selected a 'success'. *Simple sampling* is random sampling with the further provision that the probability p of success is the same at each trial. Thus p is a constant in the process; and the probability of success at any trial is independent of the success or failure of preceding trials. The value of p is the relative frequency of the occurrence of the attribute in the population from which the sample is drawn. Hence, for the sampling to be simple, either the population must be very large, or the individual selected must be returned to the population before the next trial, success or failure having been noted.

The problem connected with the drawing of a simple sample of n members is thus identical with that of a series of n independent trials, with constant probability p of success; and the results of §§ 11 and 17 are applicable. The probabilities of 0, 1, 2, ... successes in a simple sample of n members are thus the terms of the binomial expansion of $(q+p)^n$. The binomial probability distribution thus determined is called the *sampling distribution* of the number of successes in the sample. The expected value, or mean value, of the number of successes is therefore np; the variance is npq, and the standard deviation is $\sqrt{(npq)}$. This s.d. is usually called the *standard error* (s.e.) of the number of successes in a sample of size n, the deviation from the expected value np being looked upon as 'error'. The proportion of successes in a sample is obtained by dividing the number of successes by n. The expected value of the proportion of successes is therefore p; and the s.d. of the proportion of successes is $1/n$ times that of the number of successes, i.e. $\sqrt{(pq/n)}$. Thus

$$\text{s.e. of the number of successes} = \sqrt{(npq)},$$
$$\text{s.e. of the proportion of successes} = \sqrt{(pq/n)}.$$

The *precision* of the proportion of successes observed in the sample is regarded as inversely proportional to the s.e. of this proportion.

Hence the precision of the observed proportion varies as $\sqrt{n}$. In particular, to double the precision it is necessary to increase the size of the sample four-fold.

45. Large samples. Test of significance

As we have just seen, the sampling distributions of the number and the proportion of successes in a simple sample of size n are binomial distributions. It was also shown in Chapter III that, for large values of n, the binomial distribution approximates to a normal distribution, in the sense that the probabilities for corresponding intervals in the two distributions tend to equality as n increases indefinitely. Now we know that the probability that a random value of a normal variate, of S.D. σ, will lie outside the interval which extends 3σ on each side of the mean is only 0·0027. Similarly, the probability that the value will deviate from the mean by more than 2σ is 0·0456, or about $4\frac{1}{2}$ %. We may therefore conclude that, for large values of n, the probability that the number of successes in a simple sample of n members will differ from the mean by more than three times the S.E. is also very small; and that a deviation of more than twice the S.E. is rather unusual.

Bearing this in mind we have a test of the credibility of the hypothesis that a given large sample, of n members, was obtained by simple sampling from a population in which the relative frequency of the occurrence of the attribute considered is p. Suppose it is found that the number of successes in the sample differs from the expected value np by more than $3\sqrt{(npq)}$. Then an event has happened which, on the hypothesis of simple sampling, is very improbable. We conclude then that the truth of this hypothesis is itself very improbable, and we say that the difference is highly *significant*. Considerations of the aspects of the problem will then lead us to suspect either that the value of p employed is incorrect, or else that the conditions of simple sampling were not observed. A deviation from the mean less than twice the S.E. is regarded as not significant. For deviations greater than twice the S.E. the significance increases with the deviation. The dividing line between significance and non-significance is, of course, not sharply defined. But significance is usually regarded as beginning where the probability of a larger deviation is less than 5 %. It may be remarked that, while the above test may furnish evidence against the hypothesis, it cannot prove the hypothesis to be correct. The most it can do in its favour is to provide no evidence against it.

The above argument still holds if, in place of the number of successes and its s.e., we employ the proportion of successes in the sample and its s.e. The expected value of this proportion in simple sampling is p; and we compare the deviation of the actual proportion, from this value, with the s.e., $\sqrt{(pq/n)}$, of this proportion. In some cases the value of p in the population is not known, but must be estimated from the sample. The estimate obtained from a large sample may be used without serious error in place of the true value, since the s.e. of the proportion of successes is small when n is large.

Example. A certain cubical die was thrown 9,000 times, and a 5 or a 6 was obtained 3,240 times. On the assumption of random throwing, do the data indicate an unbiased die?

On the hypothesis of an unbiased die the chance of throwing a 5 or a 6 is 1/3. Thus $p = 1/3$ and $q = 2/3$. The expected number of successes is therefore 3,000, and the deviation of the actual number from this value is 240. The s.e., ϵ, of the number of successes is

$$\epsilon = \sqrt{(npq)} = \sqrt{(9000 \times \tfrac{1}{3} \times \tfrac{2}{3})} = 10\sqrt{20} = 44 \cdot 72.$$

The deviation 240 is nearly 5·4 times this s.e.; and it is therefore most unlikely to appear as a result of simple sampling with $p = 1/3$. We therefore conclude that the die is almost certainly biased, and that p is not equal to 1/3.

The estimate of p obtained from the sample is 3,240/9,000 = 0·36. The s.e. of the proportion of successes is then

$$\epsilon' = \sqrt{(0 \cdot 36 \times 0 \cdot 64/9,000)} = 0 \cdot 0050,$$

and $3\epsilon' = 0 \cdot 015$. It is therefore most unlikely that the true value of p lies outside the range $0 \cdot 36 \pm 0 \cdot 015$. In other words, the true value of p almost certainly lies between 0·345 and 0·375.

46. Comparison of large samples

Let two populations, P_1 and P_2, be tested for the prevalence of a certain attribute, by taking from them large simple samples of n_1 and n_2 members respectively; and let p_1 and p_2 be the observed proportions of successes in the samples. Is the difference, $p_1 - p_2$, significant of a real difference between the two populations with respect to the given attribute? On the hypothesis that the populations are similar in this respect, we may combine the samples to estimate the common value of the relative frequency of the occurrence of the attribute in the populations. This estimate is then

$$p = \frac{n_1 p_1 + n_2 p_2}{n_1 + n_2}. \tag{1}$$

The s.e.'s of the proportions of successes in samples of n_1 and n_2 members are $\sqrt{(pq/n_1)}$ and $\sqrt{(pq/n_2)}$ respectively; and, since the samples are independent, the variance ϵ^2 of the difference of these proportions is given by

$$\epsilon^2 = pq\left(\frac{1}{n_1} + \frac{1}{n_2}\right) \tag{2}$$

in consequence of the theorem proved in §§ 10 and 15.

On the assumption that the populations are similar, the expected value of the difference $p_1 - p_2$ is zero, for

$$E(p_1 - p_2) = E(p_1) - E(p_2) = 0.$$

The sampling distributions of p_1 and p_2 are approximately normal, when n_1 and n_2 are large; and the same is true of their difference since the samples are independent. Thus the distribution of $p_1 - p_2$ is approximately normal, with mean zero and s.d. ϵ. The probability that, in simple sampling, the difference $p_1 - p_2$ will be numerically greater than 3ϵ is therefore very small. The probability that it will be greater than 2ϵ is in the neighbourhood of 5 %; and any value smaller than this is regarded as not significant, i.e. as providing no evidence against the hypothesis.

Example 1. In a simple sample of 600 men from a certain large city, 400 are found to be smokers. In one of 900 from another large city, 450 are smokers. Do the data indicate that the cities are significantly different with respect to the prevalence of smoking among men?

Here $p_1 = 2/3$, $p_2 = 1/2$, so that $p_1 - p_2 = 1/6$. On the assumption that the cities are alike with respect to the prevalence of smoking among men, we have as our estimate of the common value of p,

$$p = \frac{850}{1500} = \frac{17}{30},$$

and the variance of the difference of the proportions for the two samples is

$$\epsilon^2 = pq\left(\frac{1}{n_1} + \frac{1}{n_2}\right) = \frac{17}{30} \times \frac{13}{30}\left(\frac{1}{600} + \frac{1}{900}\right) = 0.000682.$$

E

Hence $\epsilon = 0.026$. The observed difference is greater than 6ϵ, and is therefore highly significant. Our assumption that the populations are similar is therefore almost certainly wrong.

Next suppose that simple samples, of n_1 and n_2 members, are drawn from populations in which the proportions are p_1 and p_2 respectively ($p_1 > p_2$). Is it likely that the proportions p_1' and p_2' in the samples will be such that $p_1' - p_2' \leqslant 0$; in other words, is the real difference between the populations likely to be hidden in sampling? As before, the distribution of $p_1' - p_2'$ is approximately normal for large values of n_1 and n_2; but the mean of the distribution is now $p_1 - p_2$, and its variance, being the sum of the variances of p_1' and p_2', is given by

$$\epsilon^2 = \frac{p_1 q_1}{n_1} + \frac{p_2 q_2}{n_2}. \tag{3}$$

In order that the sample value of $p_1' - p_2'$ should be negative, its deviation from the mean must be *on the negative side*, and numerically greater than $p_1 - p_2$. If $p_1 - p_2 > 3\epsilon$, the probability of this is very small. If, however, $p_1 - p_2$ is much less than 2ϵ, such an event would not be very unusual. For the particular case in which $p_1 - p_2 = 2\epsilon$, the probability of the event is in or near the interval $2-2\frac{1}{2}\%$.

Example 2. In two large populations there are 35 and 30 % of fair-haired people. Is the difference likely to be revealed by simple samples of 1,500 and 1,000 respectively from the two populations?

Here $p_1 = 0.35$ and $p_2 = 0.30$, so that $p_1 - p_2 = 0.05$. The variance of the difference of the proportions in the samples is

$$\epsilon^2 = \frac{(0.35)(0.65)}{1500} + \frac{(0.3)(0.7)}{1000} = 0.000362,$$

so that $\epsilon = 0.019$. The difference $p_1 - p_2$ is about 2.6ϵ. The probability that the real difference between the populations will be hidden is approximately the probability that, for a random value of a normal variate, the deviation from the mean will be on the negative side and greater than 2.6 times the s.d. Since this is less than $\frac{1}{2}\%$, it is unlikely that the difference will be hidden.

47. Poissonian and Lexian sampling. Samples of varying size

Consider next a few modifications of the condition of simple sampling, beginning with *Poisson's series of trials* (cf. § 11, Ex. 3). Suppose that, in drawing a sample of attributes of n members, the chance of success changes at each drawing. Let p_i be the probability of success at the ith drawing. Then the expected value of the number of successes in the sample is the sum of the expectations at the individual drawings; and this is

$$\sum p_i = np,$$

where p is the mean of the quantities p_i. Also, since the drawings are independent, the variance ϵ^2 of the number of successes in the sample is the sum of the variances of the numbers of successes at the separate drawings, so that

$$\epsilon^2 = \sum p_i q_i = \sum p_i - \sum p_i^2.$$

If σ_p^2 is the variance of the quantities p_i, we may write this

$$\epsilon^2 = np - n(p^2 + \sigma_p^2) = npq - n\sigma_p^2. \tag{4}$$

Thus the variance of the number of successes is less than when the probability remains constant and equal to p. Dividing by n^2 we have the variance ϵ'^2 of the proportion of successes in a Poissonian sample

$$\epsilon'^2 = \frac{pq}{n} - \frac{\sigma_p^2}{n}. \tag{5}$$

Consider next a *Lexian series of trials*.* Suppose that, in taking N simple samples of attributes of n members each, the probability of success varies from one sample to another. Let p_i be the value in the ith sample $(i = 1, 2, ..., N)$. We wish to find the s.e. of the number of successes per sample, when the records of all the samples are pooled. Let p be the mean value of the probability, so that $Np = \sum p_i$. Then the expected value of the number of successes in the whole series of samples is equal to the sum of the expected values for the individual samples, and this is

$$\sum np_i = nNp.$$

Hence the expected value of the number of successes per sample is np. To find the variance ϵ^2 of the number of successes per sample, we observe that the mean square deviation from np in the ith sample is

$$np_i q_i + (np_i - np)^2.$$

Summing for all the N samples, and equating to $N\epsilon^2$, we have

$$N\epsilon^2 = n\sum p_i q_i + n^2 \sum (p_i - p)^2.$$

---

* Studied by W. Lexis in 1877.

Now the last term has the value $n^2 N\sigma_p^2$, where σ_p^2 is the variance of the quantities p_i. In the other sum we may write

$$\sum p_i q_i = \sum p_i - \sum p_i^2 = Np - N(\sigma_p^2 + p^2)$$
$$= Npq - N\sigma_p^2.$$

Substituting these values in the equation, and dividing by N, we obtain

$$\epsilon^2 = npq + n(n-1)\sigma_p^2, \tag{6}$$

which is the required variance of the number of successes per sample. Dividing by n^2 we have the variance of the proportion of successes per sample, viz.

$$\epsilon'^2 = \frac{pq}{n} + \frac{n-1}{n}\sigma_p^2. \tag{7}$$

Both the variances (6) and (7) are greater than in the case of simple sampling with constant probability p.

Lastly, we may consider the modification of simple sampling in which the probability p of success remains constant, but *the size n of the sample varies* about a mean $\bar{n}$ with variance σ_n^2. To find the variance of the number of successes per sample, we observe first that the mean number of successes per sample is $\bar{n}p$. Then, for samples of size n, the mean square deviation of the number of successes from np is npq. Consequently the mean square deviation from the general mean $\bar{n}p$ is

$$npq + (np - \bar{n}p)^2.$$

The expected value of this mean square is the required variance of the number of successes, so that

$$\epsilon^2 = \bar{n}pq + p^2\sigma_n^2. \tag{8}$$

We shall make use of this result in a later chapter.

SAMPLING OF VALUES OF A VARIABLE

48. Random and simple sampling

We pass now to the consideration of sampling of values of a variable and measurable quantity, such as height, age, yield of grain, etc. Each member of the population of individuals, objects or experiments provides a value of the variable; and we thus have a population of values of the variable, and the frequency distribution determined by it. In drawing a sample of n members from the population we are choosing n values of the variable from those of the distribution.

We have already defined *random sampling* as sampling in which each member of the population has the same chance of being chosen. In the case of a population of discrete values of the variable x, the value x_i may occur f_i times. There are then f_i members of the population each equal to x_i. Hence the probability of the value x_i in the selection of an individual by random sampling is f_i/N, which is the relative frequency of that value in the population. Similarly, if the population of values of x has a continuous distribution with relative frequency density $f(x)$, the probability that, in the random selection of an individual, the value of the variable will fall in the interval dx is $f(x)\,dx$. *Simple sampling* is random sampling with the further provision that, in the selection of an individual from the population, the probability of obtaining a value of the variable within any specified range remains constant throughout the sampling. In particular, with a population of discrete values of the variable, the probability of obtaining a specified value x_i remains constant during the sampling. Thus, in simple sampling, the system of probabilities associated with any drawing is independent of the results of preceding drawings.

A population, whose distribution is continuous, contains an infinite number of values in any finite interval in the range of the variable. In drawing a finite random sample from such a population, the probability associated with any interval remains unchanged, and the sampling is therefore simple. Thus a finite random sample from a population whose distribution is continuous is a simple sample. It is the common practice to refer to such a sample as a 'random sample'. If, however, the population contains only a limited number of values, the sampling will not be simple unless each value selected is returned to the population before the drawing of the next value.

49. Sampling distributions. Standard errors

The distribution of the variable in the population has its mean, variance, moments of higher order, partition values, etc., which are spoken of generally as the *parameters* of the population. Similarly, each simple sample from the population determines a frequency distribution of the variable, from which the mean, variance, etc., of the sample may be calculated. These may be regarded as estimates of the values of the parameters of the population. Any such estimate obtained from the sample is called a *statistic*. More generally any function of the sample values, used as an estimate of a parameter of the population, is called a statistic.

When the distribution of the variable x in the population is known, it is theoretically possible to determine the probability that the estimate z of any parameter, obtained from a simple sample of n members, will lie in the interval dz. Let this probability be denoted by $\phi(z)\,dz$. Then the density $\phi(z)$ determines a probability distribution called the *sampling distribution* of that statistic for simple samples of size n. Thus the sampling distribution is a continuous distribution, determined by the nature of the population and the size of the sample; and the S.D. of the sampling distribution is called the *standard error* (S.E.) of that statistic for samples of n members. The sampling distribution is often defined as the distribution of the values of the statistic obtained from an infinite (or very large) number of simple samples of the given size. This alternative way of looking at it may be a help to the student. The two definitions bear the same relation to each other as the *à priori* definition of probability and the empirical definition. The reader should bear in mind that a sampling distribution is essentially a probability distribution.

Distributions of statistics, for random samples from a normal population, will play a prominent part in later chapters. It was shown in § 22 that the distribution of the means of such samples is normal; and we know its S.D. In general, for normal populations, or populations whose frequency curves are unimodal and only moderately skew, the sampling distributions of many of the common statistics approximate to the normal type as n increases indefinitely; so that, for large samples, they possess the property that the probability of a sample value of the statistic deviating from its mean by more than three times its S.E. is very small. This enables us to apply the test of § 45 to statistics obtained from large samples; and the determination of the S.E. in such cases is a matter of importance. Tests appropriate to small samples will be considered in Chapter X.

50. Sampling distribution of the mean

The distribution of the means of random samples of given size, from a specified population, is a question of considerable importance. Let the sampled population have mean μ and variance σ^2. We shall first prove by elementary methods that the sampling distribution of the mean $\bar{x}$, of random samples of size n, has μ for its mean and σ^2/n for its variance. We shall then prove more generally how all the moments of the distribution of the mean may be deduced simply from those of the population, by means of the properties of the cumulative function.

Reasoning in terms of a population of discrete values, let the relative frequency of the value x_i in the population be p_i $(i = 1, 2, ..., k)$. Then

$$\mu = \sum p_i x_i \tag{9}$$

and
$$\sigma^2 = \sum p_i (x_i - \mu)^2. \tag{10}$$

In the case of a sample of *one* member, the distribution of the mean is clearly the distribution of the variable in the population. For the single value x_s in the sample, associated with probability p_s, is the mean of the sample. The expected value of the mean of the sample is therefore

$$E(x_s) = \sum p_i x_i = \mu.$$

Consequently the variance of the distribution of the mean of a sample of one is

$$E(x_s - \mu)^2 = \sum p_i (x_i - \mu)^2 = \sigma^2$$

as stated. For a sample of n values x_j $(j = 1, 2, ..., n)$, the mean $\bar{x}$ is given by

$$n\bar{x} = \sum x_j.$$

Taking the expected value of each member we have

$$E(n\bar{x}) = E(\sum x_j) = \sum E(x_j) = \sum \mu = n\mu,$$

so that
$$E(\bar{x}) = \mu. \tag{11}$$

Thus *the mean of the population is the mean of the sampling distribution of $\bar{x}$.* Further, since the values in the sample are independent, the sampling variance of their sum, $n\bar{x}$, is the sum of the variances of the separate values. But each of these values is a sample of one member, whose distribution has variance σ^2. Consequently the variance of $n\bar{x}$ is $n\sigma^2$, and its s.d. is $\sigma\sqrt{n}$. The s.d. of $\bar{x}$ is therefore $\sigma/\sqrt{n}$. This is the required s.e. of the mean.

If the distribution of values in the population is continuous, with relative frequency density $f(x)$, the relative frequency for the interval dx is $f(x)\,dx$; and this is the probability of a random value coming from that interval. To adapt the above argument we have only to replace p_i by $f(x)\,dx$, x_i by x, and summation with respect to i by integration extending over the values in the population. Thus (9) and (10) become

$$\mu = \int xf(x)\,dx$$

and

$$\sigma^2 = \int (x-\mu)^2 f(x)\,dx.$$

Summation with respect to j remains unaltered, since it extends only over the n values of the sample.

Example. A sample of 900 members is found to have a mean of 3·4 cm. Could it be reasonably regarded as a simple sample from a large population, whose mean is 3·25 cm. and s.d. 2·61 cm.?

The s.e. of the mean of a simple sample of 900 from such a population is $2\cdot61/30 = 0\cdot087$ cm. The deviation of the mean of the sample from that of the population is 0·15 cm. This deviation is less than twice the s.e. of the mean, and is therefore not significant. We conclude that the given sample might be one drawn from the population specified.

The various moments of the sampling distribution of $\bar{x}$ may be found very simply* by using the additive property of cumulants, proved in §16. For, since $n\bar{x} = \sum x_j$, it follows from this property that the cumulative function of $n\bar{x}$, being the sum of those of the variates x_j, is given by

$$K(t;\, n\bar{x}) = nK(t;\, x)$$
$$= n\kappa_1 t + n\kappa_2 t^2/2! + n\kappa_3 t^3/3! + \dots,$$

the cumulants κ_r being those of the population, and the second argument in brackets denoting the variate whose cumulative function is indicated. But the rth moment of $\bar{x}$ is obtained from that

* Cf. Fisher, 1929, 1, p. 202.

of $n\bar{x}$ by dividing by n^r. Hence the cumulative function of $\bar{x}$ is found by substituting t/n for t in the above expansion. The cumulative function of the distribution of the mean is therefore

$$K(t;\bar{x}) = \kappa_1 t + \frac{\kappa_2}{n}\frac{t^2}{2!} + \frac{\kappa_3}{n^2}\frac{t^3}{3!} + \frac{\kappa_4}{n^3}\frac{t^4}{4!} + \dots$$

The rth cumulant for the distribution of the mean of the sample is thus found from that of the population by dividing by n^{r-1}. In particular, the mean of the distribution of $\bar{x}$ is κ_1, and is therefore the same as the mean of the population. The variance of $\bar{x}$ is σ^2/n, and the third moment about the mean of the distribution is μ_3/n^2.

That the sampling distribution of the mean is approximately normal for large samples, when the population has only moderate skewness and excess, also follows from the results of § 16. For, since the rth cumulant of the distribution of the mean is obtained from that of the population by dividing by n^{r-1}, the skewness of the distribution of the mean is

$$\frac{\kappa_3}{n^2}\left(\frac{n}{\kappa_2}\right)^{3/2} = \frac{1}{\sqrt{n}} \text{ (skewness of the population)}.$$

Similarly, the excess of kurtosis of the distribution of the mean

$$\frac{\kappa_4}{n^3}\frac{n^2}{\kappa_2^2} = \frac{1}{n} \text{ (excess of population)}.$$

Thus, for large samples from a population of moderate skewness and excess, the skewness and excess of the distribution of $\bar{x}$ are small, and the distribution is approximately normal.

51. Normal population. Fiducial limits for unknown mean

Suppose that the population, from which the random sample of n values is drawn, is a normal population with mean μ and S.D. σ. Then the sample values are independent normal variates, with the same mean and S.D.; and, by the theorem of § 22, their mean $\bar{x}$ is normally distributed with mean μ and S.D. $\sigma/\sqrt{n}$. If we know σ^2 but not μ, there is a range of possible values of μ for which the observed mean $\bar{x}$ of the sample is not significant at any specified level of probability. In the sampling distribution of the mean, the relative deviation of $\bar{x}$ from its expected value is the ratio of $\bar{x} - \mu$ to the s.e. of $\bar{x}$; and this ratio is $(\bar{x} - \mu)\sqrt{n}/\sigma$. If then the observed value $\bar{x}$ is not significant at the 5 % level of probability, this relative deviation must be less than that which, in a normal distribution, is exceeded numerically with a probability of 5 %. Such a relative deviation is 1·96. Consequently the observed mean $\bar{x}$ will be not significant provided

$$\left| \frac{(\bar{x} - \mu)\sqrt{n}}{\sigma} \right| < 1·96,$$

which requires
$$\bar{x} - 1·96\sigma/\sqrt{n} < \mu < \bar{x} + 1·96\sigma/\sqrt{n}.$$

The values $\bar{x} \mp 1·96\sigma/\sqrt{n}$ are called the 95 % *fiducial limits*, or confidence limits, for the mean of the population corresponding to the given sample. They are the limits within which μ must lie, in order that the observed sample mean should not be significant at the prescribed level of probability.

Similarly, we define the fiducial limits for other levels of probability. Thus, in a normal distribution, the relative deviations which are exceeded numerically with probabilities of 2 and 1 % are 2·33 and 2·58 respectively. Hence the 98 % fiducial limits for the mean of the normal population, corresponding to the given sample, are $\bar{x} \mp 2·33\sigma/\sqrt{n}$; and the 99 % fiducial limits are $\bar{x} \mp 2·58\sigma/\sqrt{n}$.

Example. Show that, in the example of the preceding section, if the population is normal but its mean unknown, the 95 % fiducial limits for the mean are 3·23 and 3·57 cm., and the 98 % fiducial limits 3·20 and 3·60 cm.

52. Comparison of the means of two large samples

Given two independent simple samples, of n_1 and n_2 members respectively, we may wish to examine whether the difference of their means may be accounted for by fluctuations of sampling, the two samples being regarded as drawn from the same population of s.d. σ. The s.e.'s of the means of samples of n_1 and n_2 members from this population are $\sigma/\sqrt{n_1}$ and $\sigma/\sqrt{n_2}$ respectively. Hence, on the assumption that the samples are independent and drawn from this population, the s.e. ϵ of the difference of their means is given by

$$\epsilon^2 = \sigma^2 \left(\frac{1}{n_1} + \frac{1}{n_2} \right). \tag{12}$$

The sampling distribution of this difference has zero for its mean, and is approximately normal if n_1 and n_2 are large. Consequently, if the observed difference of the means exceeds 3ϵ, it can hardly be ascribed to fluctuations of sampling; and our assumption that the samples were drawn from the same population is almost certainly incorrect. If the difference is greater than 2ϵ, it is regarded as significant at the 5 % level of probability. When the variance of the population is not known, it may be estimated from the combined sample of $n_1 + n_2$ members, unless the variances of the two samples are inconsistent with the assumption that they were drawn from the same population (see § 60 below).

If the two samples are known to have come from different populations, with variances σ_1^2 and σ_2^2 respectively, we can test by a similar procedure whether the two populations may have the same mean. The standard errors of the means of samples of n_1 and n_2 members from the two populations are $\sigma_1/\sqrt{n_1}$ and $\sigma_2/\sqrt{n_2}$ respectively; and the s.e. ϵ of their difference is then given by

$$\epsilon^2 = \frac{\sigma_1^2}{n_1} + \frac{\sigma_2^2}{n_2}. \tag{13}$$

On the assumption that the two populations have the same mean, the distribution of the difference of the means of the samples has zero for its expected value, and is approximately normal for large samples. We may therefore test the significance of the difference of the sample means, by comparing it with ϵ in the usual manner. As before, if the variances of the populations are not known, they may be estimated from the large samples.

Example. A simple sample of heights of 6,400 Englishmen has a mean of 67·85 in. and a s.d. of 2·56 in., while a simple sample of heights of 1,600 Australians has a mean of 68·55 and a s.d. of 2·52 in. Do the data indicate that Australians are on the average taller than Englishmen?

The s.e. of the mean of a sample of heights of 6,400 Englishmen is 2·56/80 = 0·032 in., and that of the mean of a sample of heights of 1,600 Australians is 2·52/40 = 0·063 in. The s.e. of the difference of the means is

$$\epsilon = \sqrt{[(0{\cdot}032)^2 + (0{\cdot}063)^2]} = \sqrt{(0{\cdot}004993)} = 0{\cdot}07 \text{ in.}$$

The observed difference between the means of the samples is 0·70 in., which is 10 times its s.e. Hence the data are inconsistent with the assumption that the means of the two populations are equal; and we conclude that Australians are on the average taller than Englishmen.

53. Standard error of a partition value

Consider the S.E. of a partition value for a large random sample of n members, drawn from a continuous population in which the relative frequency density of the variable is $f(x)$. Let the partition value be that for which p is the fraction of the frequency lying above it, and q the fraction below it. Also let x_p be the partition value for the population, and $x_p + \delta x_p$ that for a large sample of n items. The relative frequency of values above x_p in the population is p. In the sample the relative frequency above $x_p + \delta x_p$ is p, and above x_p it is $p + \delta p$. Thus δp is the relative frequency in the sample for the interval δx_p. Now for a large sample δx_p and δp are small; and δp is, to within infinitesimals of higher order, the relative frequency for the interval δx_p in the population. Consequently

$$\delta p = f(x_p)\,\delta x_p = y\,\delta x_p,$$

where y is the ordinate at x_p for the frequency curve $y = f(x)$ of the population. On squaring we have

$$(\delta p)^2 = y^2 (\delta x_p)^2.$$

Now y is independent of the sample, and δx_p and δp vary about zero as mean. Hence, on taking expected values of each side, we obtain

$$E(\delta x_p)^2 = \frac{1}{y^2} E(\delta p)^2.$$

Now $E(\delta x_p)^2$ is the variance of the sampling distribution of x_p, and its square root is the S.E. of this partition value. Similarly, $E(\delta p)^2$ is the variance of the proportion of successes in n trials, with constant probability p that the value of the variable selected will be greater than x_p; and we know that this variance is pq/n. Consequently, on taking the square root of both sides of the above equation, we obtain the required S.E. ϵ of the partition value* as

$$\epsilon = \frac{1}{y}\sqrt{\frac{pq}{n}} = \frac{1}{f(x_p)}\sqrt{\frac{pq}{n}}. \tag{14}$$

If the value of y for the population is not given, it may be estimated from the frequency distribution of the large sample.

* See also Kendall, 1940, 4.

An important case is that in which the distribution of the variable in the population is *normal*, with s.d. σ. The value of p is the area under the normal curve to the right of the partition value; and the table of areas in § 21 enables us to read off the value of x_p/σ. The table of ordinates for the normal curve in § 20 then gives the corresponding value of σy; and substitution of the values of y, p, q and n in (14) gives the required s.e. For example, in the case of the median

$$p = \tfrac{1}{2}, \quad q = \tfrac{1}{2}, \quad x/\sigma = 0, \quad \sigma y = 0\cdot3989.$$

Hence the s.e. *of the median*, for a large sample of n members from a normal population, is

$$\frac{\sigma}{0\cdot3989}\sqrt{\frac{\tfrac{1}{2}\times\tfrac{1}{2}}{n}} = 1\cdot25\frac{\sigma}{\sqrt{n}}.$$

This is 25 % greater than the s.e. of the mean. For the upper quartile $p = \tfrac{1}{4}$, the area from the mean to the quartile is $0\cdot25$, giving

$$x/\sigma = 0\cdot6745, \quad \sigma y = 0\cdot3178.$$

The s.e. of a quartile is therefore

$$\frac{\sigma}{0\cdot3178}\sqrt{\frac{\tfrac{1}{4}\times\tfrac{3}{4}}{n}} = 1\cdot36\frac{\sigma}{\sqrt{n}}.$$

Example. Prove in the same manner that, for large samples of n members from a normal population of s.d. σ,

s.e. of 1st and 9th deciles $= 1\cdot71\sigma/\sqrt{n}$,

s.e. of 2nd and 8th deciles $= 1\cdot43\sigma/\sqrt{n}$,

s.e. of 3rd and 7th deciles $= 1\cdot32\sigma/\sqrt{n}$,

s.e. of 4th and 6th deciles $= 1\cdot27\sigma/\sqrt{n}$.

COLLATERAL READING

Yule and Kendall, 1937, 1, chapters xviii–xx.

Rietz, 1927, 2, pp. 114–18 and 146–52.

Rietz (ed.), 1924, 1, chapters v and vi.

Aitken, 1939, 1, chapter iii.

Kenney, 1939, 3, part ii, chapter vi.

Rider, 1939, 6, §§ 34–37.

Mills, 1938, 1, pp. 452–73.

Camp, 1934, 1, part ii, chapter iv (to p. 254).

Ezekiel, 1930, 2, chapter ii.

Tippett, 1931, 2, chapter iii.

Snedecor, 1938, 3, chapters iii, iv and viii.

Jones, 1924, 3, chapter xii.

EXAMPLES VI

1. A biased coin was thrown 400 times, and heads resulted 240 times. Find the s.e. of the proportion of heads in 400 throws; and deduce that the probability of throwing heads in a single trial almost certainly lies between 0·53 and 0·67.

2. A random sample of 500 pineapples was taken from a large consignment, and 65 were found to be bad. Show that the s.e. of the proportion of bad ones in a sample of this size is 0·015; and deduce that the percentage of bad pineapples in the consignment almost certainly lies between 8·5 and 17·5.

3. In a random sample of 800 adults from the population of a certain large city, 600 are found to have dark hair. In a random sample of 1,000 adults from the inhabitants of another large city, 700 are dark-haired. Show that the difference of the proportions of dark-haired people is nearly 2·4 times the s.e. of this difference for samples of the above sizes.

4. In two large populations there are 30 and 25 % respectively of fair-haired people. Is this difference likely to be hidden in samples of 1,200 and 900 respectively from the two populations? (The difference, 0·05, in the proportions is more than $2\frac{1}{2}$ times the s.e. of this difference for such samples. Hence it is unlikely that the real difference will be hidden.)

5. Given that, on the average, 4 % of insured men of age 65 die within a year, and that 60 of a particular group of 1,000 such men died within a year, show that this group cannot be regarded as a representative sample, seeing that the actual deviation of the proportion of deaths is more than three times the s.e. of the proportion for samples of this size.

6. In sampling of attributes a sample is drawn containing an even number n of members. In drawing each of the first $\frac{1}{2}n$ members the probability of success is p, and for each of the remaining ones it is $1 - p$, the drawings being independent. Show that the expected number of successes is $\frac{1}{2}n$, and the variance of the number of successes is npq.

7. The sampling distribution of a certain statistic is normal, with mean 2·0 and s.e. 1·5. Show that the probability that a simple sample will yield a value of the statistic greater than 5·0 is 0·02275; and the probability that the value found will lie outside the range −1·0 to 5·0 is 0·0455.

Also find c if the probability of getting a value of the statistic greater than c is 0·07. (*Ans. c* = 4·213.)

8. A normal population has a mean of 0·1 and a s.d. of 2·1. Find the probability that the mean of a simple sample of 900 members will be negative. (*Ans.* 0·077 nearly.)

9. A sample of 900 members is found to have a mean of 3·47 cm. Can it be reasonably regarded as a simple sample from a large population with mean 3·23 cm. and s.d. 2·31 cm.? (*Ans.* No. The deviation $\bar{x} - \mu$ is more than three times the s.e. of the mean.)

10. The means of simple samples of 1,000 and 2,000 are 67·5 and 68·0 in. respectively. Can the samples be regarded as drawn from the same population of s.d. 2·5 in.? (*Ans.* No. The difference of the means is more than 5 times the s.e. of the difference, which is 0·097 in. nearly.)

11. Show that the s.e. of the 8th decile of a simple sample of 900 members drawn from a normal population of s.d. 2·5 cm. is 0·12 cm. nearly.

12. Prove that the coefficient of correlation between errors in the partition values corresponding to proportions p and p' of the frequency lying above them is, for large samples from a continuous population, $\sqrt{(p'q/pq')}$ in which $p > p'$. In particular, for the lower and upper quartiles this coefficient is 1/3. (Cf. Yule and Kendall, 1937, 1, p. 385.)

13. Using the value $\epsilon = 1·363\sigma/\sqrt{n}$ for the s.e. of the lower (or upper) quartile in a large sample of n values from a normal population of s.d. σ, find the s.e. of the semi-interquartile range.

Since the interquartile range is the difference between the upper and lower quartiles, and the correlation between errors in these statistics in a large sample is $1/3$ (cf. Ex. 12), the s.e. of the inter-quartile range is, in virtue of § 32 (46),

$$\sqrt{(\epsilon^2 + \epsilon^2 - \tfrac{2}{3}\epsilon^2)} = 2\epsilon/\sqrt{3}.$$

The s.e. of the semi-interquartile range is therefore

$$\epsilon/\sqrt{3} = 1\cdot 363\sigma/\sqrt{(3n)} = 0\cdot 787\sigma/\sqrt{n}.$$

14. The quantities x_i $(i = 1, 2, ..., n)$ are independent values chosen at random, one from each of n populations with the same mean, and with variances σ_i^2. Show that the linear estimate of the common mean, which has the least sampling variance, is that obtained by weighting the x's inversely as their variances. Also prove that this minimum variance is H/n, where H is the harmonic mean of the variances σ_i^2. (Cf. Ex. IV, 5.)

15. From a finite population of N values, with variance σ^2, a random sample of n values is drawn without replacements. Show that the sampling variance of the mean of the sample is

$$(N-n)\sigma^2/n(N-1).$$

16. *Variable Poissonian sampling.* The three types of sampling considered in § 47 are all included in the following. Suppose that, in the drawing of a Poissonian sample, there are various types of sampling, each type having its own probability. Let ϖ_k be the probability of the kth type of Poissonian sample, n_k the size of a sample of this type, p_{kj} $(j = 1, ..., n_k)$, the probabilities of success at the individual drawings in this type, so that the expected number of successes in such a sample is $\bar{x}_k = \sum_j p_{kj}$. In virtue of § 47 (4) the variance of the number of successes per sample in this type is

$$\epsilon_k^2 = \bar{x}_k - \frac{1}{n_k}\bar{x}_k^2 - n_k\sigma_k^2,$$

where σ_k^2 is the variance of the p_{kj} for the kth type of sampling. The expected number of successes per sample when all types are possible is

$$\bar{x} = \sum_k \varpi_k\bar{x}_k,$$

and the variance of the number of successes per sample for all types, being the mean square deviation of the number of successes per sample from $\bar{x}$, is given by

$$\epsilon^2 = \sum_k \varpi_k [\epsilon_k^2 + (\bar{x}_k - \bar{x})^2]$$

$$= \sum_k \varpi_k \left[\bar{x}_k - \frac{1}{n_k} \bar{x}_k^2 - n_k \sigma_k^2 \right] + \sigma_{\bar{x}}^2,$$

where $\sigma_{\bar{x}}^2$ is the variance of the $\bar{x}_k$ for all types. Hence the *general formula**

$$\epsilon^2 = \bar{x} + \sigma_{\bar{x}}^2 - \sum_k \varpi_k \left[\frac{1}{n_k} \bar{x}_k^2 + n_k \sigma_k^2 \right].$$

In the particular case, for example, in which all the p_{kj} have a common value, p, we have

$$\sigma_k = 0, \quad \bar{x}_k = pn_k, \quad \bar{x} = p\bar{n}, \quad \sigma_{\bar{x}}^2 = p^2 \sigma_n^2,$$

and the above formula becomes

$$\epsilon^2 = p\bar{n} + p^2 \sigma_n^2 - p^2 \sum_k \varpi_k n_k = \bar{n}pq + p^2 \sigma_n^2,$$

which is § 47 (8).

17. In a certain population the proportion of members possessing a given characteristic is p. Prove that, if p may be varied, the probability of obtaining m such members in a simple sample of n from the population is greatest when $p = m/n$.

18. Prove that, in simple sampling from a population, the expected value of the rth moment of the sample about any fixed value is equal to the rth moment of the population about that value.

* Cf. Aitken, 1939, 1, pp. 53–4 and Coolidge, 1925, 2, pp. 66–72.

CHAPTER VII

STANDARD ERRORS OF STATISTICS

54. Notation. Variances of population and sample

Having already considered the s.e.'s of the mean and the partition values, we now pass to those of various other statistics. We shall argue in terms of a population of discrete values x_i ($i = 1, 2, ..., k$), the relative frequency of the value x_i in the population being p_i. The argument covers approximately the case in which the values of the population are grouped in k classes of small class interval, the ith class being centred at the value x_i, and having a relative frequency p_i. The sum of the relative frequencies of all the other classes is therefore q_i, where as usual $p_i + q_i = 1$. Then, as in § 50, the mean μ of the population is given by

$$\mu = \sum_{i=1}^{k} p_i x_i \tag{1}$$

and its variance σ^2 by

$$\sigma^2 = \sum p_i (x_i - \mu)^2 = \sum p_i x_i^2 - \mu^2. \tag{2}$$

Consider now a simple sample of n members drawn from this population. The sample values fall into the same classes as above; that is to say, the values x_i are the same for the sample as for the population, but the relative frequencies of the classes vary with the sample, fluctuating about the corresponding values for the population. The probability that, in simple sampling, a value drawn will belong to the ith class is p_i. Hence the frequency f_i of that class, in a sample of n members, has mean value np_i, since this is the expected value of the number of successes in n independent trials with constant probability p_i of success. Thus

$$E(f_i) = np_i. \tag{3}$$

The mean $\bar{x}$ of the sample is given by

$$n\bar{x} = \sum f_i x_i, \tag{4}$$

and its variance S^2 by

$$nS^2 = \sum f_i (x_i - \bar{x})^2 = \sum f_i x_i^2 - n\bar{x}^2. \tag{5}$$

We may notice at the outset that the expected value of S^2 in sampling is not σ^2 but $(n-1)\sigma^2/n$. To prove this we observe that S^2, being the variance of the sample, is the second moment of the sample about μ, diminished by $(\bar{x}-\mu)^2$. Thus

$$S^2 = \frac{1}{n}\Sigma f_i(x_i-\mu)^2 - (\bar{x}-\mu)^2.$$

Consider the expectation of both members of this equation. In the first term on the right $(x_i-\mu)$ remains constant during sampling; and therefore, in virtue of (3), the expected value of this term is

$$\frac{1}{n}\Sigma\left[(x_i-\mu)^2 E(f_i)\right] = \Sigma p_i(x_i-\mu)^2 = \sigma^2.$$

Further, $E(\bar{x}-\mu)^2$ is the sampling variance of the mean, and is therefore equal to σ^2/n. Substituting these values we obtain

$$E(S^2) = \sigma^2 - \frac{\sigma^2}{n} = \frac{n-1}{n}\sigma^2 \tag{6}$$

as stated above. If then we write

$$s^2 = \frac{n}{n-1}S^2 = \frac{1}{n-1}\Sigma f_i(x_i-\bar{x})^2, \tag{7}$$

our result is $\qquad\qquad E(s^2) = \sigma^2. \tag{8}$

We say that s^2 is an *unbiased* estimate of σ^2, because its mean value in sampling is σ^2. It is in this sense a better estimate of the population variance than S^2. Incidentally also it follows from the above argument that the second moment of the sample, about the mean of the population, has σ^2 for its mean value.

55. Standard errors of class frequencies

We have seen that, in simple sampling from the above population, the probability that a value will fall into the ith class of the sample is p_i, and that the frequency f_i of this class has mean value np_i. The deviation of this frequency from its mean value will be denoted by δf_i. The s.e. of f_i is the square root of the expected value of $(\delta f_i)^2$; and, since f_i is the number of successes in n trials of constant probability p_i, its variance is np_iq_i. Thus

$$E(\delta f_i)^2 = np_iq_i = np_i(1-p_i). \tag{9}$$

If p_i is unknown, an estimate of its value derived from the sample is f_i/n; and, as an approximation to the above sampling variance, we have

$$E(\delta f_i)^2 \sim f_i(1 - f_i/n). \tag{10}$$

But this is a fair approximation only if n is large. A better estimate is obtained by multiplying this expression by $n/(n-1)$. To prove this we consider the expected value of the second member of (10). Since $E(f_i^2)$ is the second moment of f_i about zero in its sampling distribution, we have

$$E(f_i^2) = \text{sampling variance of } f_i + [E(f_i)]^2$$
$$= np_i q_i + (np_i)^2.$$

Consequently

$$E(f_i - f_i^2/n) = np_i - [p_i(1 - p_i) + np_i^2]$$
$$= (n-1)\, p_i(1 - p_i) = \frac{n-1}{n}\, E(\delta f_i)^2.$$

Hence the formula

$$E(\delta f_i)^2 \sim \frac{n}{n-1} f_i(1 - f_i/n) \tag{10'}$$

gives a better estimate than (10), since the expected value of the second member of (10') is equal to the first member, so that it is an unbiased estimate of the sampling variance of f_i. For a large sample the factor $n/(n-1)$ is nearly equal to unity, and (10) gives a sufficiently good estimate.

56. Covariance of the frequencies in different classes

Since the sum of the class frequencies in any sample is constant, for samples of given size n, it follows that

$$\sum_i \delta f_i = 0. \tag{11}$$

The covariance between the frequencies f_i and f_j of the ith and jth classes is the covariance, $E(\delta f_i \delta f_j)$, of their deviations δf_i and δf_j from their mean values np_i and np_j. The calculation of this covariance may be associated with a correlation table in the values of δf_i and δf_j. Consider the array in which δf_i has a fixed value. Bearing (11) in mind we assume that, in all the samples for which

δf_i has this fixed value, the excess δf_i produces an equal deficiency which is distributed among the other class frequencies, on the average, in proportion to their expected values; that is to say, we assume that, in the array with constant δf_i, the average value of δf_j is

$$\overline{\delta f_j} = - \left(\frac{p_j}{1-p_i} \right) \delta f_i. \tag{12}$$

In calculating the covariance we may, as proved in § 33, replace each value of δf_j in the array by the mean value $\overline{\delta f_j}$ for that array. Then

$$E(\delta f_i \delta f_j) = - E\left(\frac{p_j(\delta f_i)^2}{1-p_i} \right) = - \frac{p_j}{1-p_i} E(\delta f_i)^2$$
$$= - n p_i p_j \tag{13}$$

in virtue of (9). This is the required covariance.

If the values of p_i and p_j are unknown, an approximation is obtained by taking them as f_i/n and f_j/n. The second member of (13) is then replaced by $-f_i f_j/n$. A better estimate, however, is given by

$$E(\delta f_i \delta f_j) \sim - \frac{f_i f_j}{n-1}. \tag{14}$$

This may be shown by determining the expected value of the second member. Thus, in virtue of § 23 (5),

$$E(\delta f_i \delta f_j) = E(f_i f_j) - E(f_i) E(f_j),$$

and therefore, by (13) and (3),

$$E(f_i f_j) = n^2 p_i p_j - n p_i p_j = n(n-1) p_i p_j$$
$$= - (n-1) E(\delta f_i \delta f_j).$$

Consequently, (14) gives an unbiased estimate of the covariance, since the expected value of the second member is equal to the covariance. In the case of a large sample, the denominator $n-1$ may be treated as n.

57. Standard errors in moments about a fixed value

It is customary to employ Greek symbols for moments of the population, and the corresponding Italics for those of the sample. Thus μ_r and μ_r' denote respectively the rth moment of the population about the mean, and about some other specified value, while m_r and m_r' denote the corresponding moments of the sample. For the rth moment of the sample about $x = 0$ we have

$$nm_r' = \sum f_i x_i^r.$$

Hence, owing to deviations δf_i of the class frequencies from their mean values np_i, we have a deviation $\delta m_r'$ of the rth moment from its mean value, given by

$$n\delta m_r' = \sum x_i^r \delta f_i.$$

On squaring both sides we obtain

$$n^2(\delta m_r')^2 = \sum x_i^{2r}(\delta f_i)^2 + \sum_{i,j}' x_i^r x_j^r \delta f_i \delta f_j,$$

where $\sum_{i,j}'$ denotes summation over all integral values of i and j from 1 to k, except those for which $i = j$. Taking expected values of both members of this equation we deduce, since the values x_i are the same for each sample,

$$n^2 E(\delta m_r')^2 = \sum x_i^{2r} np_i(1-p_i) - \sum_{i,j}' x_i^r x_j^r np_i p_j.$$

Consequently, on dividing by n and rearranging,

$$nE(\delta m_r')^2 = \sum p_i x_i^{2r} - \sum p_i^2 x_i^{2r} - \sum_{i,j}' p_i p_j x_i^r x_j^r$$
$$= \mu_{2r}' - (\sum p_i x_i^r)^2 = \mu_{2r}' - (\mu_r')^2,$$

the moments μ_{2r}' and μ_r' being those of the population. Hence the required formula

$$E(\delta m_r')^2 = \frac{1}{n}(\mu_{2r}' - \mu_r'^2). \tag{15}$$

If, however, on taking expected values as above we use the estimates (10′) and (14), we obtain the unbiased estimate of the sampling variance of m_r',

$$E(\delta m_r')^2 \sim \frac{1}{n-1}(m_{2r}' - m_r'^2) \tag{16}$$

in terms of the moments of the sample.

Since the mean of a distribution is the first moment about $x = 0$, we may deduce the *variance of the mean* of a sample of n members by putting $r = 1$ in the above. Thus (15) becomes

$$E(\delta \bar{x})^2 = \frac{1}{n}(\mu_2' - \mu^2) = \frac{\sigma^2}{n}, \tag{17}$$

and (16) gives, in terms of sample moments,

$$E(\delta \bar{x})^2 \sim \frac{1}{n-1}(m_2' - \bar{x}^2) = \frac{S^2}{n-1} = \frac{s^2}{n}, \tag{18}$$

in agreement with (8).

58. Covariance of moments of different orders about a fixed value

The calculation of the covariance of the qth and rth moments of the sample about the fixed value $x = 0$ is similar to the above. With the same notation we have

$$nm_q' = \sum f_i x_i^q, \qquad nm_r' = \sum f_i x_i^r,$$

and therefore $\qquad n\delta m_q' = \sum x_i^q \delta f_i, \qquad n\delta m_r' = \sum x_i^r \delta f_i.$

On multiplying corresponding sides of these equations we obtain

$$n^2 \delta m_q' \delta m_r' = \sum_i x_i^{q+r}(\delta f_i)^2 + {\sum_{i,j}}' x_i^q x_j^r \delta f_i \delta f_j.$$

Now take the expected value of each member of the equation. Then, in virtue of (9) and (13),

$$n^2 E(\delta m_q' \delta m_r') = \sum_i x_i^{q+r} n p_i (1 - p_i) - {\sum_{i,j}}' x_i^q x_j^r n p_i p_j,$$

and consequently, on dividing by n and rearranging,

$$nE(\delta m_q' \delta m_r') = \sum_i p_i x_i^{q+r} - \sum_i p_i^2 x_i^{q+r} - {\sum_{i,j}}' p_i p_j x_i^q x_j^r$$

$$= \mu_{q+r}' - (\sum_i p_i x_i^q)(\sum_j p_j x_j^r) = \mu_{q+r}' - \mu_q' \mu_r'.$$

Hence the required formula

$$E(\delta m_q' \delta m_r') = \frac{1}{n}(\mu_{q+r}' - \mu_q' \mu_r') \tag{19}$$

in terms of the moments of the population. If, however, on taking expected values as above, we use the approximations (10') and (14), we obtain the unbiased estimate

$$E(\delta m_q' \delta m_r') \sim \frac{1}{n-1}(m_{q+r}' - m_q' m_r') \tag{20}$$

in terms of moments of the sample.

59. Standard errors of the variance and the standard deviation of a large sample

The mean of a sample changes with the sample. Hence we cannot use (15) to obtain the sampling variance of the second moment about the mean of the sample. In the case of large samples we may obtain the required result as follows. The variance m_2 of the sample is connected with the second moment m_2' about $x = 0$ by the usual relation

$$m_2 = m_2' - \bar{x}^2.$$

Consequently

$$\delta m_2 = \delta m_2' - 2\bar{x}\,\delta\bar{x},$$

and therefore, on squaring,

$$(\delta m_2)^2 = (\delta m_2')^2 - 4\bar{x}\,\delta\bar{x}\,\delta m_2' + 4\bar{x}^2(\delta\bar{x})^2, \tag{i}$$

where

$$\delta\bar{x} = \frac{1}{n}\sum_i x_i\,\delta f_i, \quad \delta m_2' = \frac{1}{n}\sum_i x_i^2\,\delta f_i. \tag{ii}$$

Suppose now that the origin of x is taken to coincide with the mean of the population. Then $\bar{x}$ becomes identical with $\delta\bar{x}$, for it is now the deviation of the mean of the sample from the mean of the population. If we take expected values of both members of (i) we may show that, when n is large, the contributions of the second and third terms on the right are small compared with that of the first term. By (15), since the origin is now the mean of the population, the expected value of $(\delta m_2')^2$ is $(\mu_4 - \mu_2^2)/n$. Also $\bar{x}^2(\delta\bar{x})^2$ is $(\delta\bar{x})^4$, and its expected value is of the order $1/n^2$, in virtue of (17). Similarly, $\bar{x}\,\delta\bar{x}\,\delta m_2'$ is $(\delta\bar{x})^2\,\delta m_2'$, and its expected value is of the order $1/n^{3/2}$, in virtue of (15) and (17). If then n is large, the expectation of the first term on the right of (i) is large compared with those of the second and third terms, and we have the approximate formula*

$$E(\delta m_2)^2 = \frac{1}{n}(\mu_4 - \mu_2^2). \tag{21}$$

---

* Fisher has shown that, for samples of any size, the exact sampling variance of s^2 (the unbiased estimate of population variance) is

$$(\mu_4 - 3\mu_2^2)/n + 2\mu_2^2/(n-1) \quad \text{(Fisher, 1929, 1, p. 206)}.$$

For a normal population this expression has the value $2\sigma^4/(n-1)$.

The s.e. of the variance of the sample is the square root of this quantity. In the particular case of a *normal* population, with s.d. σ, we have $\mu_4 = 3\sigma^4$ and $\mu_2 = \sigma^2$; so that, for large samples from a normal population,

$$E(\delta m_2)^2 = 2\sigma^4/n. \tag{22}$$

From (21) may be deduced the s.e. of the s.d. of large samples. For the variance of the sample we have, with the above notation, $m_2 = S^2$, and therefore

$$\delta m_2 = 2S\,\delta S.$$

For large samples the factor S may be taken as equal to the population parameter σ, δS being small. Hence on squaring, and taking expected values of both members, we have

$$E(\delta m_2)^2 = 4\sigma^2 E(\delta S)^2,$$

and therefore, in virtue of (21),

$$E(\delta S)^2 = \frac{\mu_4 - \mu_2^2}{4\sigma^2 n}. \tag{23}$$

In the case of large samples from a normal population this is simply

$$E(\delta S)^2 = \frac{\sigma^2}{2n}, \tag{24}$$

and the s.e. of the s.d. is, in this case, $\sigma/\sqrt{(2n)}$.

60. Comparison of the standard deviations of two large samples

For two independent large samples of n_1 and n_2 members from the same population, the variance ϵ^2 of the difference of their standard deviations, being the sum of the variances of their standard deviations, is in virtue of (23),

$$\epsilon^2 = \frac{\mu_4 - \mu_2^2}{4\mu_2}\left(\frac{1}{n_1} + \frac{1}{n_2}\right). \tag{25}$$

In particular, if the population is normal, this equation takes the simple form

$$\epsilon^2 = \tfrac{1}{2}\sigma^2\left(\frac{1}{n_1} + \frac{1}{n_2}\right). \tag{26}$$

The expected value of the difference of the s.d.'s of two such samples is zero. Hence, by comparing the actual difference $S_1 - S_2$

with the S.E. ϵ, we may test in the usual manner the credibility of the hypothesis that the samples were drawn from the same population.

For two large simple samples from different populations, whose moments are μ_4, μ_2 and $\tilde{\mu}_4$, $\tilde{\mu}_2$ respectively, the S.E. of the difference of their standard deviations is given by

$$\epsilon^2 = \frac{\mu_4 - \mu_2^2}{4\mu_2 n_1} + \frac{\tilde{\mu}_4 - \tilde{\mu}_2^2}{4\tilde{\mu}_2 n_2}. \tag{27}$$

And in particular, if the populations are normal with standard deviations σ_1 and σ_2 respectively,

$$\epsilon^2 = \frac{\sigma_1^2}{2n_1} + \frac{\sigma_2^2}{2n_2}. \tag{28}$$

Example. The S.D. of a simple sample of 2,000 members is 5·9 years, and that of an independent sample of 2,500 members is 6·1 years. May the samples be reasonably regarded as from the same normal population?

On the hypothesis that they are from the same normal population, an approximate value for its S.D. is 6 years, and the S.E. of the difference of the standard deviations of samples of the above sizes is, by (26),

$$\epsilon = 6\sqrt{(\tfrac{1}{4000} + \tfrac{1}{5000})} = 6(0\cdot0212) = 0\cdot127 \text{ year.}$$

The actual difference of 0·2 year is less than 1·6ϵ, and is therefore not significant at the 5 % level of probability. There is thus no real evidence against the hypothesis.

SAMPLING FROM A BIVARIATE POPULATION

61. Sampling covariance of the means of the variables

Suppose now that the population is bivariate, with p_i as relative frequency of the pair of values (x_i, y_i), or of the class with centre at that point. The means of x and y in the population are then

$$\mu = \sum_i p_i x_i, \quad \mu' = \sum_i p_i y_i,$$

and the moments of orders q and r in x and y respectively are

$$\mu'_{q,r} = \sum p_i x_i^q y_i^r, \quad \mu_{q,r} = \sum p_i (x_i - \mu)^q (y_i - \mu')^r,$$

the first about the point $(0, 0)$, and the second about the mean of the population. In particular $\mu_{1,1}$ is the covariance of x and y in the population.

In a simple sample of n pairs of values, let f_i be the frequency of the ith class. Then the values of the moments of the sample are given by

$$nm'_{q,r} = \Sigma f_i x_i^q y_i^r, \quad nm_{q,r} = \Sigma f_i (x_i - \bar{x})^q (y_i - \bar{y})^r,$$

the covariance of x and y in the sample being $m_{1,1}$. The argument of §§ 55 and 56 still holds, and the formulae

$$E(\delta f_i)^2 = np_i(1-p_i), \quad E(\delta f_i \delta f_j) = -np_i p_j$$

are still valid.

Corresponding to (17) we may now prove that the *covariance of the means* $\bar{x}, \bar{y}$, in samples from a bivariate population, is $\mu_{1,1}/n$. For

$$n\bar{x} = \Sigma f_i x_i, \quad n\bar{y} = \Sigma f_i y_i.$$

Hence the deviations $\delta \bar{x}, \delta \bar{y}$ from μ, μ' respectively, and the deviations δf_i from their mean values np_i, are connected by

$$n\,\delta \bar{x} = \Sigma x_i \delta f_i, \quad n\,\delta \bar{y} = \Sigma y_i \delta f_i,$$

and therefore on multiplication

$$n^2 \delta \bar{x}\, \delta \bar{y} = \sum_i x_i y_i (\delta f_i)^2 + \sum_{i,j}' x_i y_j \delta f_i \delta f_j.$$

Taking the expected value of each member we deduce

$$n^2 E(\delta \bar{x}\, \delta \bar{y}) = \sum_i x_i y_i np_i(1-p_i) - \sum_{i,j}' x_i y_j np_i p_j,$$

and therefore

$$nE(\delta \bar{x}\, \delta \bar{y}) = \sum_i p_i x_i y_i - \left(\sum_i p_i x_i\right)\left(\sum_j p_j y_j\right)$$
$$= \mu'_{1,1} - \mu\mu' = \mu_{1,1}.$$

Hence the required result

$$E(\delta \bar{x}\, \delta \bar{y}) = \mu_{1,1}/n. \tag{29}$$

From this it follows immediately that the coefficient of correlation between $\bar{x}$ and $\bar{y}$ is equal to the correlation ρ between x and y in the population. For the correlation between $\bar{x}$ and $\bar{y}$ is

$$\frac{\text{covariance of } \bar{x} \text{ and } \bar{y}}{(\text{s.e. of } \bar{x})\,(\text{s.e. of } \bar{y})} = \frac{\mu_{1,1}/n}{(\sigma_1/\sqrt{n})(\sigma_2/\sqrt{n})} = \frac{\mu_{1,1}}{\sigma_1 \sigma_2} = \rho,$$

σ_1^2 and σ_2^2 being the variances of x and y in the population.

We may also prove that the expected value of the covariance of x and y in the sample is given by

$$E(m_{1,1}) = \frac{n-1}{n}\mu_{1,1} \tag{30}$$

which corresponds to (6). For, by the known relation, §23 (5),

$$m_{1,1} = \frac{1}{n}\sum f_i(x_i - \mu)(y_i - \mu') - (\bar{x} - \mu)(\bar{y} - \mu').$$

Taking the expected value of each member we have, in virtue of (29),

$$E(m_{1,1}) = \frac{1}{n}\sum (x_i - \mu)(y_i - \mu')E(f_i) - \frac{1}{n}\mu_{1,1}$$

$$= \sum p_i(x_i - \mu)(y_i - \mu') - \frac{1}{n}\mu_{1,1}$$

$$= \left(1 - \frac{1}{n}\right)\mu_{1,1}$$

as required.

62. Variance and covariance of moments about a fixed point

The moment of the sample about the fixed point $(0,0)$, of order q in x and r in y, is given by

$$nm'_{q,r} = \sum_i x_i^q y_i^r f_i,$$

and therefore, with the usual notation,

$$n\,\delta m'_{q,r} = \sum_i x_i^q y_i^r \delta f_i.$$

Squaring both sides, and taking expected values as in §57, we deduce

$$n^2 E(\delta m'_{q,r})^2 = \sum_i x_i^{2q} y_i^{2r} np_i(1 - p_i) - \sum_{i,j}' x_i^q y_i^r x_j^q y_j^r np_i p_j$$

$$= n\mu'_{2q,2r} - n\mu'^2_{q,r},$$

and therefore

$$E(\delta m'_{q,r})^2 = \frac{1}{n}(\mu'_{2q,2r} - \mu'^2_{q,r}) \tag{31}$$

in terms of the moments of the population. This is the sampling variance of $m'_{q,r}$. If, in taking expected values, we use the approximations (10′) and (14), we obtain the unbiased estimate

$$E(\delta m'_{q,r})^2 \sim \frac{1}{n-1}(m'_{2q,2r} - m'^2_{q,r}) \tag{32}$$

in terms of the moments of the sample.

Similarly, we may find the sampling covariance of the moments $m'_{q,r}$ and $m'_{s,t}$ about $(0, 0)$, by multiplying together the expressions for $\delta m'_{q,r}$ and $\delta m'_{s,t}$ and taking expected values of both sides. Proceeding as above we obtain the result

$$E(\delta m'_{q,r}\, \delta m'_{s,t}) = \frac{1}{n}\,(\mu'_{q+s,\,r+t} - \mu'_{q,r}\mu'_{s,t}). \tag{33}$$

In particular by putting $r = s = 0$ and $q = t = 1$, and observing that $\mu'_{1,0} = \mu$, $\mu'_{0,1} = \mu'$, we find again the covariance of the means

$$E(\delta \bar{x}\, \delta \bar{y}) = \frac{1}{n}\,(\mu'_{1,1} - \mu\mu') = \frac{1}{n}\,\mu_{1,1}.$$

As in other cases we may deduce the unbiased estimate corresponding to (33), with sample moments and denominator $n-1$.

63. Standard error of the covariance of a large sample

By argument similar to that of § 59 we may find the s.e. of the covariance of a large sample from a bivariate population. For, with the usual notation, since

$$m_{1,1} = m'_{1,1} - \bar{x}\bar{y},$$

we have $\quad\quad\quad \delta m_{1,1} = \delta m'_{1,1} - \bar{y}\,\delta \bar{x} - \bar{x}\,\delta \bar{y}.$

If now we take the origin at the mean of the population, $\bar{x}$ becomes identical with $\delta \bar{x}$ and $\bar{y}$ with $\delta \bar{y}$. Consequently, if we retain only the principal terms we have, on squaring the last equation and taking expected values,

$$E(\delta m_{1,1})^2 = E(\delta m'_{1,1})^2,$$

and, since the origin is at the mean of the population, this is, in virtue of (31),

$$E(\delta m_{1,1})^2 = \frac{1}{n}\,(\mu_{2,2} - \mu_{1,1}^2), \tag{34}$$

giving the square of the s.e. of the covariance of the sample.

64. Standard error of the coefficient of correlation

The s.e. of the coefficient of correlation, r, for large samples of n pairs from a bivariate population, may be found as follows.*

* Cf. Bowley, 1920, 2, pp. 422–3.

Omitting the comma between the two subscripts in the moment symbols we have, by the usual formula,

$$r = m_{11}/\sqrt{(m_{20}m_{02})},$$

and therefore, by logarithmic differentiation,

$$\frac{\delta r}{r} = \frac{\delta m_{11}}{m_{11}} - \frac{1}{2}\frac{\delta m_{20}}{m_{20}} - \frac{1}{2}\frac{\delta m_{02}}{m_{02}}. \tag{i}$$

If the origin is taken at the mean of the population, so that $\bar{x}$ and $\bar{y}$ are identical with $\delta\bar{x}$ and $\delta\bar{y}$, and we retain only small quantities of the first order, we find as in the preceding section

$$\delta m_{11} = \delta m'_{11} = \frac{1}{n}\sum_i x_i y_i \delta f_i.$$

Similarly, from the relations

$$m_{20} = m'_{20} - \bar{x}^2, \quad m_{02} = m'_{02} - \bar{y}^2,$$

we find, on retaining only the principal terms,

$$\delta m_{20} = \delta m'_{20} = \frac{1}{n}\sum x_i^2 \delta f_i,$$

$$\delta m_{02} = \delta m'_{02} = \frac{1}{n}\sum y_i^2 \delta f_i.$$

Now substitute these values in (i). Then, since n is large and we are retaining only small quantities of the first order, we may replace the denominators in (i) by the moments and correlation coefficient ρ for the population. Thus

$$\frac{\delta r}{\rho} = \sum\left(\frac{x_i y_i}{\mu_{11}} - \frac{1}{2}\frac{x_i^2}{\mu_{20}} - \frac{1}{2}\frac{y_i^2}{\mu_{02}}\right)\frac{\delta f_i}{n}$$

$$= \sum F(x_i, y_i)\,\delta f_i/n,$$

where $F(x_i, y_i)$ denotes the expression in brackets. Squaring both sides we have

$$\left(\frac{\delta r}{\rho}\right)^2 = \frac{1}{n^2}\sum_i [F(x_i, y_i)\,\delta f_i]^2 + \frac{1}{n^2}\sum'_{i,j} F(x_i, y_i) F(x_j, y_j)\,\delta f_i \delta f_j,$$

and therefore, on taking expected values,

$$E(\delta r)^2 = \frac{\rho^2}{n}\left[\sum F^2(x_i, y_i)\,p_i(1-p_i) - \sum'_{i,j} F(x_i, y_i) F(x_j, y_j)\,p_i p_j\right]$$

$$= \frac{\rho^2}{n}\left[\sum F^2(x_i, y_i)\,p_i - \left\{\sum_i F(x_i, y_i)\,p_i\right\}^2\right].$$

Now the second sum on the right is equal to zero. For

$$\Sigma F(x_i, y_i) p_i = \Sigma \left(\frac{x_i y_i}{\mu_{11}} - \frac{x_i^2}{2\mu_{20}} - \frac{y_i^2}{2\mu_{02}} \right) p_i = 1 - \tfrac{1}{2} - \tfrac{1}{2} = 0.$$

On substituting the value of $F^2(x_i, y_i)$ we deduce the sampling variance of r in the form

$$E(\delta r)^2 = \frac{\rho^2}{n} \left[\frac{\mu_{22}}{\mu_{11}^2} + \frac{\mu_{40}}{4\mu_{20}^2} + \frac{\mu_{04}}{4\mu_{02}^2} - \frac{\mu_{31}}{\mu_{20}\mu_{11}} - \frac{\mu_{13}}{\mu_{02}\mu_{11}} + \frac{\mu_{22}}{2\mu_{20}\mu_{02}} \right]. \quad (35)$$

This result assumes a very simple form in the case of a bivariate normal population. The parameter values are then

$$\mu_{22} = (1 + 2\rho^2) \sigma_1^2 \sigma_2^2, \quad \mu_{11} = \rho \sigma_1 \sigma_2, \quad \mu_{20} = \sigma_1^2, \quad \mu_{02} = \sigma_2^2,$$

$$\mu_{40} = 3\sigma_1^4, \quad \mu_{04} = 3\sigma_2^4, \quad \mu_{31} = 3\rho\sigma_1^3\sigma_2, \quad \mu_{13} = 3\rho\sigma_1\sigma_2^3,$$

and on substituting these values we obtain

$$E(\delta r)^2 = \frac{1}{n} (1 - \rho^2)^2.$$

Consequently $\quad\quad\quad$ s.e. of $r = \dfrac{1 - \rho^2}{\sqrt{n}}.$ $\quad\quad\quad\quad$ (36)

These formulae hold for large values of n. Their application, however, is limited by the fact that the sampling distribution of r is not even approximately normal when r is fairly large. The s.e. test for r should therefore be applied only for large samples and moderate values of r. A much more useful test, which is applicable to small or large samples, will be considered in Chapter x.

COLLATERAL READING

RIETZ, 1927, 2, chapter v.
YULE and KENDALL, 1937, 1, chapter xxi.
JONES, 1924, 3, chapters xiii and xiv.
RIETZ (ed.), 1924, 1, chapter v.
CAMP, 1934, 1, part ii, chapter iv.
BOWLEY, 1920, 2, part ii, chapter ix.
KENDALL, 1943, 2, chapter ix.

EXAMPLES VII

1. In a sample of 10,000 from a normal population the S.D. is 2·52 cm. Show that the S.E. of this quantity is 0·018 cm. nearly, and hence that the S.D. of the population almost certainly lies between 2·46 and 2·58 cm.

2. A random sample of 6,400 members from a certain population has a S.D. of 6·80 years, and a fourth moment of 3298 yr.[4] Show that three times the S.E. of the S.D. is 0·094 nearly, and hence that the S.D. of the population almost certainly lies between 6·7 and 6·9 years.

3. The S.D. of a random sample of 1,000 members is 5·9 years, and that of an independent sample of 900 members is 6·1 years. Show that the samples may reasonably be regarded as drawn from equally variable normal populations, since the difference of their standard deviations is very little greater than its S.E.

4. *Standard error of the coefficient of variation.* By definition this coefficient is $V = S/\bar{x} = \sqrt{m_2}/\bar{x}$. Logarithmic differentiation gives

$$\frac{\delta V}{V} = \frac{\delta m_2}{2m_2} - \frac{\delta \bar{x}}{\bar{x}}.$$

Squaring both sides and taking expected values show that, for large samples

$$E(\delta V)^2 = V^2 \left[\frac{\mu_4 - \mu_2^2}{4n\mu_2^2} + \frac{\mu_2}{n\bar{x}^2} - \frac{1}{\bar{x}\mu_2} E(\delta \bar{x}\, \delta m_2) \right].$$

It can be shown that, for samples from a normal population, $E(\delta \bar{x}\, \delta m_2) = 0$. Show that in this case the above result leads to

$$\text{s.e. of } V = V \sqrt{[(1 + 2V^2)/2n]}.$$

5. *Standard error of a moment about the mean of a large sample.* Proceeding as in § 59 we have

$$nm_q = \sum (x_i - \bar{x})^q f_i$$
$$= \sum x_i^q f_i - q\bar{x} \sum x_i^{q-1} f_i + \dots,$$

and therefore

$$\delta m_q = \delta m'_q - q\, \delta \bar{x} m'_{q-1} - q\bar{x}\, \delta m'_{q-1} + \dots.$$

If the origin is taken at the mean of the population, $\bar{x}$ becomes identical with $\delta\bar{x}$; and, as we need retain only the terms of lowest order, we neglect those after the first two. On squaring we may write the result

$$(\delta m_q)^2 = (\delta m_q')^2 + q^2 \mu_{q-1}'^2 (\delta\bar{x})^2 - 2q\mu_{q-1}' \delta\bar{x} \delta m_q',$$

and on taking expected values of both sides we obtain, in virtue of (15) and (19), the origin being the mean of the population,

$$E(\delta m_q)^2 = \frac{1}{n} (\mu_{2q} - \mu_q^2 + q^2 \mu_{q-1}^2 \sigma^2 - 2q\mu_{q+1}\mu_{q-1}).$$

The square root of this quantity is the s.e. of m_q.

6. Deduce that, for large samples from a normal population of s.d. σ, the s.e.'s of m_3 and m_4 are $\sigma^3\sqrt{(6/n)}$ and $\sigma^4\sqrt{(96/n)}$ respectively.

7. A random sample of 3,600 pairs of values from a bivariate normal population showed a correlation coefficient of 0·45. Find limits to the correlation in the population.

Since the sample is large we may take 0·45 as an approximate value for ρ. The s.e. of r is then $(1 - 0·2025)/60 = 0·0133$, so that $3\epsilon = 0·04$ nearly. The coefficient of correlation in the population almost certainly lies between 0·41 and 0·49.

8. A random sample of 2,500 pairs of values from a bivariate normal population showed a correlation of 0·1. Is this really significant of correlation in the population?

On the assumption that the population is uncorrelated we have the s.e. of $r = 1/50 = 0·02$. The actual value is 5 times this s.e., and is therefore significant.

Show that the above correlation of 0·1 would not be significant in a sample of less than 400 pairs.

9. By means of the result of Examples VI, 18 (p. 129) deduce* the formulae (15) of §57 and (19) of §58.

* Cf. Kendall, 1943, 2, pp. 205–6.

F

CHAPTER VIII

BETA AND GAMMA DISTRIBUTIONS

65. Beta and Gamma Functions

For the benefit of the student who is not familiar with them, we shall first prove the elementary properties of the Beta and Gamma functions. The integral

$$\Gamma(n) = \int_0^\infty e^{-x} x^{n-1} dx \tag{1}$$

converges if n is positive. It is a function of n called the *Gamma Function*. Clearly

$$\Gamma(1) = \int_0^\infty e^{-x} dx = 1. \tag{2}$$

Also, if $n-1$ is positive, we have on integration by parts

$$\Gamma(n) = \left[-e^{-x} x^{n-1} \right]_0^\infty + \int_0^\infty (n-1) x^{n-2} e^{-x} dx,$$

so that

$$\Gamma(n) = (n-1)\,\Gamma(n-1). \tag{3}$$

Hence, if n is a positive integer,

$$\Gamma(n) = (n-1)(n-2)\ldots 2 \cdot 1 \cdot \Gamma(1) = (n-1)!. \tag{4}$$

On account of the property expressed by (3) and (4), $\Gamma(n)$ is often denoted by $(n-1)!$ whether n is integral or not. Also, on writing x^2 in place of x in (1), we have the alternative formula

$$\Gamma(n) = 2 \int_0^\infty x^{2n-1} \exp(-x^2)\, dx. \tag{5}$$

And, by an obvious substitution, it is easily verified that, if a is positive,

$$\int_0^\infty e^{-ax} x^{n-1} dx = a^{-n}\,\Gamma(n). \tag{6}$$

The integral

$$B(m, n) = \int_0^1 x^{m-1}(1-x)^{n-1}\,dx \tag{7}$$

also converges if m and n are positive. It is a function of m and n called the *Beta Function*. That it is symmetrical in m and n is easily shown by the substitution $z = 1 - x$. Then (7) becomes

$$B(m, n) = \int_0^1 z^{n-1}(1-z)^{m-1}\,dz = B(n, m).$$

Further, on substituting $x = \sin^2\theta$ in (7) we obtain

$$B(m, n) = 2\int_0^{\frac{1}{2}\pi} \sin^{2m-1}\theta \, \cos^{2n-1}\theta\,d\theta, \tag{8}$$

and therefore in particular

$$B(\tfrac{1}{2}, \tfrac{1}{2}) = 2\int_0^{\frac{1}{2}\pi} d\theta = \pi, \tag{9}$$

while from (7) it is obvious that

$$B(1, 1) = 1. \tag{10}$$

Lastly the substitution $x = 1/(1+y)$ in (7) leads to an important alternative definition, viz.

$$B(m, n) = \int_0^\infty \frac{y^{n-1}\,dy}{(1+y)^{m+n}}, \tag{11}$$

and in this integral m and n may be interchanged, in virtue of the symmetry of the function.

Example. Show that

$$B(m, n) = \int_0^1 \frac{x^{m-1} + x^{n-1}}{(1+x)^{m+n}}\,dx.$$

This may be deduced from (11) by dividing the range of integration into two parts, 0 to 1 and 1 to ∞, and putting $y = 1/x$ in the integration over the second part.

66. Relation between the two functions

That the Beta and Gamma functions are connected by the relation

$$B(m, n) = \frac{\Gamma(m)\,\Gamma(n)}{\Gamma(m+n)} \tag{12}$$

may be proved as follows. Consider the integrals

$$I_1 = 2\int_0^a x^{2m-1}\exp(-x^2)\,dx, \quad I_2 = 2\int_0^a y^{2n-1}\exp(-y^2)\,dy,$$

whose limiting values as a tends to infinity are $\Gamma(m)$ and $\Gamma(n)$ respectively, in virtue of (5). Then

$$I_1 I_2 = 4 \int_0^a dx \int_0^a \exp(-x^2 - y^2)\, x^{2m-1} y^{2n-1}\, dy,$$

or, on changing to polar coordinates,

$$I_1 I_2 = 4 \iint \exp(-r^2)\, r^{2m+2n-1} \cos^{2m-1}\theta \, \sin^{2n-1}\theta \, dr\, d\theta,$$

the integration extending over the square $OACB$ (see Fig. 4, p. 66). Since the integrand is positive, this integral is intermediate in value between the integrals of the same function extended over the quadrant OAB, of radius a, and the quadrant OPQ, of radius $a\sqrt{2}$. Hence $I_1 I_2$ lies in value between

$$4 \int_0^{\frac{1}{2}\pi} \cos^{2m-1}\theta \, \sin^{2n-1}\theta \, d\theta \int_0^a \exp(-r^2)\, r^{2m+2n-1}\, dr,$$

and the corresponding integral with 0 and $a\sqrt{2}$ as the limits for r. But, as a tends to infinity, each of these integrals tends to the limit $B(m,n)\,\Gamma(m+n)$, while I_1 and I_2 tend to the limits $\Gamma(m)$ and $\Gamma(n)$ respectively. Consequently

$$B(m,n)\,\Gamma(m+n) = \Gamma(m)\,\Gamma(n)$$

as stated in (12).

Putting $m = n = \frac{1}{2}$ in this result we have, in virtue of (2) and (9),

$$\pi = \Gamma(\tfrac{1}{2})\,\Gamma(\tfrac{1}{2}).$$

Consequently $\qquad\qquad \Gamma(\tfrac{1}{2}) = \sqrt{\pi} \qquad\qquad\qquad (13)$

or $\qquad\qquad \int_0^\infty \frac{e^{-x}}{\sqrt{x}}\, dx = \sqrt{\pi}.$

By writing x^2 in place of x in this integral, or by putting $n = \frac{1}{2}$ in (5), we deduce

$$\int_0^\infty \exp(-x^2)\, dx = \tfrac{1}{2}\sqrt{\pi}. \qquad\qquad (14)$$

Example 1. Show that, if m is a positive integer,

$$\Gamma(m+\tfrac{1}{2}) = \frac{2m-1}{2}\,\frac{2m-3}{2}\cdots \tfrac{1}{2}\sqrt{\pi}.$$

Example 2. Show that, if m and n are positive integers,

$$B(m, n) = \frac{(m-1)!\,(n-1)!}{(m+n-1)!}.$$

In particular, $B(1, n) = 1/n$.

Example 3. Show that

$$B(m+1, n)/B(m, n) = m/(m+n)$$

and $\qquad B(m+2, n)/B(m, n) = m(m+1)/(m+n)\,(m+n+1).$

Example 4. Show that

$$B(m+2, n-2)/B(m, n) = m(m+1)/(n-1)\,(n-2).$$

Example 5. Show that

$$\int_0^a (a-x)^{m-1} x^{n-1}\,dx = a^{m+n-1}\,B(m, n).$$

67. Gamma distribution and Gamma variates

In virtue of (1) a continuous variable x, which is distributed with probability density

$$\phi(x) = \frac{e^{-x} x^{l-1}}{\Gamma(l)} \tag{15}$$

throughout the range 0 to ∞, is called a *Gamma variate* with parameter l; and its distribution is a *Gamma distribution*.* The factor $1/\Gamma(l)$ ensures that the integral of $\phi(x)$ over the whole range of values of x is unity. The reader should sketch the probability curve $y = \phi(x)$ for the distribution. He will find that it is asymptotic to the x-axis and that, if $l > 1$, it has a mode at $x = l - 1$. If $l > 2$ it also touches the x-axis at the origin; while, if $1 < l < 2$, it is tangent to the y-axis at that point. If, however, $0 < l < 1$, the curve is asymptotic to both axes.

The *expected value* of the variate in the distribution is given by

$$E(x) = \int_0^\infty \frac{e^{-x} x^l}{\Gamma(l)}\,dx = \frac{\Gamma(l+1)}{\Gamma(l)} = l. \tag{16}$$

The second moment about $x = 0$ is similarly

$$\mu_2' = \int_0^\infty x^2 \phi(x)\,dx = \Gamma(l+2)/\Gamma(l) = l(l+1).$$

* This belongs to Karl Pearson's Type III. See Kendall, 1943, 2, pp. 137–43.

Hence the *variance* is given by

$$\sigma^2 = l(l+1) - l^2 = l. \tag{17}$$

Example. Show that the rth moment about $x = 0$ is

$$\mu_r' = l(l+1)(l+2) \dots (l+r-1),$$

and deduce that the third moment about the mean is $2l$.

An important example of a Gamma variate is associated with the normal distribution. If x is normally distributed with mean a and S.D. σ, the probability that a random value of the variate will fall in the interval dx is

$$dP = \frac{1}{\sigma \sqrt{(2\pi)}} \exp\left[-(x-a)^2/2\sigma^2\right] dx. \tag{18}$$

Let u be defined by

$$u = \tfrac{1}{2}(x-a)^2/\sigma^2,$$

so that, as x varies from $-\infty$ to $+\infty$, u varies from $+\infty$ to 0 and then from 0 to $+\infty$. For values of x between a and $+\infty$,

$$x - a = \sigma \sqrt{(2u)}$$

and

$$dx = \sigma\, du/\sqrt{(2u)},$$

so that

$$dP = \frac{e^{-u} u^{\frac{1}{2}-1}\, du}{2\sqrt{\pi}}.$$

But the probability that u falls in the interval du is double this, since there is an equal probability that u will fall in this interval when x lies between $-\infty$ and a. Consequently the probability differential for the variate u is

$$dp = \frac{e^{-u} u^{\frac{1}{2}-1}\, du}{\Gamma(\frac{1}{2})},$$

so that u is a Gamma variate with parameter $\frac{1}{2}$. We may therefore state

THEOREM I. *If x is normally distributed with mean a and standard deviation σ, then $\tfrac{1}{2}(x-a)^2/\sigma^2$ is a Gamma variate with parameter $\tfrac{1}{2}$.*

A Gamma variate with parameter l may be referred to briefly as a $\gamma(l)$ variate, the symbol being used adjectively.*

---

* The objection to describing it as a $\Gamma(l)$ variate is the additional meaning thus given to the symbol $\Gamma(l)$.

68. Sum of independent Gamma variates

The moments of the Gamma distribution, and the distribution of the sum of independent Gamma variates, may be deduced from the moment generating function. The m.g.f. of a $\gamma(l)$ variate with respect to the origin is given by

$$M(t) = \int_0^\infty \frac{e^{tx}e^{-x}x^{l-1}}{\Gamma(l)}\,dx = \int_0^\infty \frac{e^{-(1-t)x}x^{l-1}}{\Gamma(l)}\,dx = \frac{1}{(1-t)^l}$$
$$= (1-t)^{-l} \quad (|t| < 1), \tag{19}$$

and the cumulative function is therefore

$$K(t) = -l\log(1-t)$$
$$= l(t + t^2/2 + t^3/3 + t^4/4 + \ldots). \tag{20}$$

Thus the mean of the distribution is l, and the variance also l, while the other cumulants are

$$\kappa_3 = 2!\,l, \quad \kappa_4 = 3!\,l, \quad \ldots, \quad \kappa_r = (r-1)!\,l. \tag{21}$$

Suppose now that x and y are independent Gamma variates with parameters l and m respectively. Then the m.g.f. of their sum, being equal to the product of their m.g.f.'s, is $(1-t)^{-(l+m)}$. But this is the m.g.f. of a $\gamma(l+m)$ variate. We thus have

Theorem II. *The sum of two independent Gamma variates, with parameters l and m, is a Gamma variate with parameter $l+m$.*

The converse of this theorem is almost equally important. It may be stated:

Theorem III. *If the sum of two independent positive variates is a Gamma variate with parameter $l+m$, and one of them is a Gamma variate with parameter l, then the other is a Gamma variate with parameter m.*

For, if $M(t)$ denotes the m.g.f. of the last variate, we have, on equating the m.g.f. of the sum to the product of those of its components,

$$(1-t)^{-(l+m)} = (1-t)^{-l}M(t),$$

whence
$$M(t) = (1-t)^{-m},$$

and the second component is therefore a $\gamma(m)$ variate.

On account of the importance of these theorems we shall prove Theorem II from first principles. Let x and y be independent $\gamma(l)$ and $\gamma(m)$ variates, and let $z = x + y$. Suppose first that y has a fixed value in the interval dy. Then $dz = dx$. The probability that a random value of x will fall in the interval dx is

$$dp = e^{-x}x^{l-1}\,dx/\Gamma(l),$$

and therefore the probability that, for a given value of y, z will lie in the interval dz is

$$dp = e^{-(z-y)}(z-y)^{l-1}\,dz/\Gamma(l).$$

But the chance that y will have a value in the interval dy is

$$dp' = e^{-y}y^{m-1}\,dy/\Gamma(m).$$

The probability that simultaneously z will lie in the interval dz and y in the interval dy is the product $dp\,dp'$. By integration we then have the probability that, for any value of y, z will lie in the interval dz as

$$dP = \frac{e^{-z}\,dz}{\Gamma(l)\,\Gamma(m)}\int_0^z (z-y)^{l-1}y^{m-1}\,dy.$$

To evaluate the integral put $y = zt$, so that $dy = z\,dt$. Since z is the sum of the positive variates x and y, it is never less than y, so that t lies within the range 0 to 1. Consequently

$$dP = \frac{e^{-z}z^{l+m-1}\,dz}{\Gamma(l)\,\Gamma(m)}\int_0^1 (1-t)^{l-1}t^{m-1}\,dt$$

$$= \frac{B(l,m)}{\Gamma(l)\,\Gamma(m)}e^{-z}z^{l+m-1}\,dz = \frac{e^{-z}z^{l+m-1}\,dz}{\Gamma(l+m)},$$

and z is therefore a $\gamma(l+m)$ variate as stated.

Repeated application of this theorem shows that the sum of n independent Gamma variates with parameters m_i $(i = 1, 2, ..., n)$ is a Gamma variate with parameter $\sum m_i$. In particular, in virtue of Theorem I, we may state

THEOREM IV. *If x_i $(i = 1, 2, ..., n)$ are n independent variates, normally distributed about a common mean zero with standard deviations σ_i, and $\chi^2 = \sum\limits_i x_i^2/\sigma_i^2$, then $\frac{1}{2}\chi^2$ is a Gamma variate with parameter $\frac{1}{2}n$.*

69. Beta distribution of the first kind

In virtue of the first definition, (7), of the Beta function we shall say that a continuous variate x, which is distributed with probability density

$$\phi(x) = \frac{x^{l-1}(1-x)^{m-1}}{B(l,m)} \tag{22}$$

throughout the range of values 0 to 1, is a *Beta variate of the first kind* with parameters l and m; and its distribution is a *Beta distribution of the first kind.** Such a variate may be referred to briefly as a $\beta_1(l, m)$ variate. The reader should sketch the probability curve for the distribution, distinguishing the different cases. He will find that, if l and m are each greater than 1, there is a modal value

$$(l-1)/(l+m-2).$$

If $l > 2$ the curve touches the x-axis at the origin; while, if $1 < l < 2$, it is tangent to the y-axis at that point. If, however, $0 < l < 1$, the curve is asymptotic to the y-axis. Similar remarks hold for the shape of the curve near $x = 1$, according to the value of m.

The *mean value* of x is given by

$$E(x) = \int_0^1 \frac{x^l(1-x)^{m-1}\,dx}{B(l,m)} = \frac{B(l+1,m)}{B(l,m)} = \frac{l}{l+m}. \tag{23}$$

The second moment μ_2' about $x = 0$ is similarly

$$\mu_2' = \frac{B(l+2,m)}{B(l,m)} = \frac{l(l+1)}{(l+m)(l+m+1)}.$$

From these it follows that the *variance* is

$$\sigma^2 = \frac{lm}{(l+m)^2(l+m+1)}. \tag{24}$$

Corresponding to Theorem II there is a fundamental theorem which may be stated:

THEOREM V. *If x and y are independent Gamma variates with parameters l and m respectively, the quotient $x/(x+y)$ is a Beta variate of the first kind with parameters l and m.*

This may be proved from first principles. If we write

$$z = \frac{x}{x+y}, \quad \text{then} \quad x = \frac{yz}{1-z},$$

---

* This belongs to Karl Pearson's Type I.

and since x and y are both positive, the range of z is from 0 to 1. Suppose first that y has a fixed value in the interval dy. Then

$$dx = y\,dz/(1-z)^2.$$

The probability that, for this value of y, x will lie in the interval dx and therefore z in the interval dz is

$$dp = \frac{e^{-x}x^{l-1}dx}{\Gamma(l)} = \frac{1}{\Gamma(l)}\left(\frac{yz}{1-z}\right)^{l-1}\exp\left(-\frac{yz}{1-z}\right)\frac{y\,dz}{(1-z)^2}.$$

But the chance that y will have a value in the interval dy is

$$dp' = \frac{e^{-y}y^{m-1}dy}{\Gamma(m)}.$$

The probability that simultaneously z will lie in the interval dz and y in the interval dy is the product $dp\,dp'$. By integration we then find the probability that, for any value of y, z will lie in the interval dz as

$$dP = \frac{z^{l-1}dz}{\Gamma(l)\,\Gamma(m)\,(1-z)^{l+1}}\int_0^\infty y^{l+m-1}\exp\left(-\frac{y}{1-z}\right)dy.$$

To evaluate the integral put $y = (1-z)\,t$. Then the limits for t are 0 and ∞, and we obtain

$$dP = \frac{z^{l-1}(1-z)^{m-1}\,\Gamma(l+m)\,dz}{\Gamma(l)\,\Gamma(m)} = \frac{z^{l-1}(1-z)^{m-1}\,dz}{B(l,m)},$$

showing that z is a $\beta_1(l,m)$ variate as stated.

We may remark in passing that, if z is a $\beta_1(l,m)$ variate and $v = 1-z$, then v is a $\beta_1(m,l)$ variate. For the range of v is also from 0 to 1, and $|dv| = |dz|$. Expressing the probability differential for z in terms of v and dv, we obtain

$$dP = v^{m-1}(1-v)^{l-1}\,dv/B(l,m),$$

which shows that v is a $\beta_1(m,l)$ variate. And the reader will observe that dP depends on the magnitude $|dv|$ of the interval dv, but not upon the sign of dv/dz.

70. Alternative proof of theorems

A combined proof of Theorems II and V may be given more simply as follows.* As before let x and y be independent $\gamma(l)$ and

***

* This proof is given by Sawkins, 1940, 3, p. 212.

$\gamma(m)$ variates respectively. Then the probability that a random value of x will fall in the interval dx and, at the same time, a random value of y will fall in the interval dy is

$$dP = \frac{1}{\Gamma(l)\,\Gamma(m)}\, e^{-(x+y)}x^{l-1}y^{m-1}\,dx\,dy.$$

Now introduce the new variables

$$u = x+y, \quad v = x/(x+y)$$

so that $\qquad\qquad x = uv, \quad y = u(1-v).$

Then as x and y range from 0 to ∞, u ranges from 0 to ∞ and v from 0 to 1. Also

$$\left|\frac{\partial(x,y)}{\partial(u,v)}\right| = u.$$

From the above expression for the probability differential dP it follows that the probability density in the joint distribution of x and y is

$$\frac{e^{-x-y}x^{l-1}y^{m-1}}{\Gamma(l)\,\Gamma(m)} = \frac{e^{-u}u^{l+m-2}v^{l-1}(1-v)^{m-1}}{\Gamma(l)\,\Gamma(m)}.$$

But the area of the element of the xy-plane bounded by the curves along which u has the values u and $u+du$ respectively, and those along which v has the values v and $v+dv$ is

$$\left|\frac{\partial(x,y)}{\partial(u,v)}\right| du\,dv = u\,du\,dv.$$

Hence the probability that, in a random selection of x and y, the representative point (x,y) will fall in this element of area is

$$dp = \frac{e^{-u}u^{l+m-1}\,du}{\Gamma(l+m)} \cdot \frac{v^{l-1}(1-v)^{m-1}\,dv}{B(l,m)}. \qquad (25)$$

Since this is the probability that simultaneously u will fall in the interval du and v in the interval dv it follows that these variates are independent, and that u is a $\gamma(l+m)$ variate and v a $\beta_1(l,m)$ variate, as stated in the above theorems.

71. Product of a $\beta_1(l, m)$ variate and a $\gamma(l+m)$ variate

We may now prove an important property, suggested by Theorem V, which may be stated:

THEOREM VI. *The product of a $\beta_1(l, m)$ variate and an independent $\gamma(l+m)$ variate is a $\gamma(l)$ variate.*

Let v be the $\beta_1(l, m)$ variate and u the $\gamma(l+m)$ variate, and let $z = uv$. The probability differential for v is

$$dp = \frac{v^{l-1}(1-v)^{m-1}\,dv}{B(l,m)}. \tag{26}$$

Hence, for a fixed value of u in the interval du, the probability that z will fall in the interval dz is

$$dp = \frac{1}{B(l,m)}\left(\frac{z}{u}\right)^{l-1}\left(1-\frac{z}{u}\right)^{m-1}\frac{dz}{u}.$$

Multiply this by the probability that a random value of u will fall in the interval du, and integrate over the range of u from z to ∞ (since $z \leqslant u$). Thus we have the probability that, for any value of u, z will lie in the interval dz as

$$dP = \frac{z^{l-1}\,dz}{B(l,m)\,\Gamma(l+m)}\int_z^\infty e^{-u}u^{l+m-1}\left(1-\frac{z}{u}\right)^{m-1}\frac{du}{u^l}$$

$$= \frac{z^{l-1}\,dz}{\Gamma(l)\,\Gamma(m)}\int_z^\infty e^{-u}(u-z)^{m-1}\,du.$$

To evaluate the integral put $u - z = zt$. Then t ranges from 0 to ∞, and we have for the probability differential of z,

$$dP = \frac{e^{-z}z^{l-1}\,dz}{\Gamma(l)\,\Gamma(m)}\int_0^\infty e^{-zt}z^m t^{m-1}\,dt = \frac{e^{-z}z^{l-1}\,dz}{\Gamma(l)}.$$

Consequently z is a $\gamma(l)$ variate as stated.

72. Beta distribution of the second kind

In agreement with the alternative definition of $B(m, n)$ contained in (11) we may define a *Beta variate of the second kind*, with positive parameters l and m, as a continuous variate x which is distributed with probability density

$$\phi(x) = \frac{x^{l-1}}{B(l,m)\,(1+x)^{l+m}} \tag{27}$$

throughout the range $x = 0$ to $x = \infty$. Such a variate may be referred to briefly as a $\beta_2(l, m)$ variate. Its distribution is a *Beta distribution* *of the second kind.** It is important to remember that, while the values of a Beta variate of the first kind are from 0 to 1, those of a Beta variate of the second kind are from 0 to ∞. The reader should sketch the different forms of the probability curve for the latter distribution. He will observe a considerable resemblance to the case of a Gamma distribution. If $l > 1$ there is a *mode* at

$$x = (l-1)/(m+1). \tag{28}$$

The curve is asymptotic to the x-axis; and, if $l > 2$, it touches this axis also at the origin. If $1 < l < 2$ the curve touches the y-axis at the origin; and if $0 < l < 1$ the curve is asymptotic to both axes.

When $m > 1$ the *mean value* of the variate is given by

$$E(x) = \frac{1}{B(l,m)} \int_0^\infty \frac{x^l\,dx}{(1+x)^{l+m}} = \frac{B(l+1, m-1)}{B(l,m)} = \frac{l}{m-1}. \tag{29}$$

Also, if $m > 2$, the second moment about $x = 0$ is

$$\mu_2' = \frac{1}{B(l,m)} \int_0^\infty \frac{x^{l+1}\,dx}{(1+x)^{l+m}} = \frac{B(l+2, m-2)}{B(l,m)} = \frac{l(l+1)}{(m-1)(m-2)}.$$

Consequently the *variance* is

$$\sigma^2 = \frac{l(l+1)}{(m-1)(m-2)} - \frac{l^2}{(m-1)^2} = \frac{l(l+m-1)}{(m-1)^2(m-2)}. \tag{30}$$

We may observe that, if x is a Beta variate of the second kind, its reciprocal is a variate of the same kind with parameters interchanged. For, if x is a $\beta_2(l, m)$ variate, its probability density is given by (27). If then we put $x = 1/y$, we have $|dx| = |dy|/y^2$; and therefore, since y lies in the interval dy when x lies in the interval dx, the probability of this is

$$dP = \frac{(1/y)^{l-1}\,(dy/y^2)}{B(l,m)\,(1 + 1/y)^{l+m}} = \frac{y^{m-1}\,dy}{B(l,m)\,(1+y)^{l+m}}.$$

Consequently y is a $\beta_2(m, l)$ variate.

---

* This belongs to Karl Pearson's Type VI.

An important relation between the two kinds of Beta variates is expressed by

THEOREM VII. *To each Beta variate of the first kind corresponds a pair of Beta variates of the second kind; and conversely.*

Let v be a $\beta_1(l, m)$ variate, so that its probability differential is

$$dP = \frac{v^{l-1}(1-v)^{m-1} dv}{B(l, m)}, \tag{31}$$

and let w be defined by

$$v = 1/(1+w) \tag{32}$$

or its equivalent $\qquad w = (1-v)/v.$

Then $|dv| = |dw|/(1+w)^2$, and by substitution we have the probability differential of w as

$$dP = \frac{w^{m-1} dw}{B(l, m)(1+w)^{l+m}}. \tag{33}$$

Since the range of w is from 0 to ∞, it follows that w is a $\beta_2(m, l)$ variate. Its reciprocal is therefore a $\beta_2(l, m)$ variate, and the first part of the theorem is proved.

Conversely, given that w is a $\beta_2(m, l)$ variate with probability differential (33), we may define a variate v by means of (32). Substitution then shows that the probability differential of v is given by (31); and since v ranges from 0 to 1, it is a $\beta_1(l, m)$ variate. The variable $1 - v$ is therefore a $\beta_1(m, l)$ variate, and the second part of the theorem is proved.

73. Quotient of independent Gamma variates

Corresponding to Theorem V, according to which a Beta variate of the first kind is determined by two independent Gamma variates, we have

THEOREM VIII. *The quotient of two independent Gamma variates, with parameters l and m, is a $\beta_2(l, m)$ variate.*

Let x and y be independent Gamma variates with parameters l and m respectively, and let $v = y/(y+x)$. Then, in virtue of Theorem V, v is a $\beta_1(m, l)$ variate. But, if w is the quotient x/y, clearly

$$v = \frac{1}{1+x/y} = \frac{1}{1+w},$$

so that w is a $\beta_2(l, m)$ variate as stated.

The theorem may also be proved by the method of § 70, which leads to the additional information that the sum $x+y$ is distributed independently of the quotient x/y. For if

$$u = x+y, \quad v = x/y$$

we have

$$\frac{\partial(u, v)}{\partial(x, y)} = -\frac{x+y}{y^2} = -\frac{(1+v)^2}{u},$$

and therefore

$$\left| \frac{\partial(x, y)}{\partial(u, v)} \right| = \frac{u}{(1+v)^2}.$$

Since the probability that x and y fall simultaneously in the intervals dx and dy is

$$dp = \frac{x^{l-1}y^{m-1}e^{-x-y}\,dx\,dy}{\Gamma(l)\,\Gamma(m)},$$

the probability density for their joint distribution is

$$\frac{x^{l-1}y^{m-1}e^{-x-y}}{\Gamma(l)\,\Gamma(m)} = \frac{e^{-u}u^{l+m-2}v^{l-1}}{\Gamma(l)\,\Gamma(m)\,(1+v)^{l+m-2}}.$$

Consequently the probability that, in a random choice of x and y, the representative point (x, y) will fall in the area bounded by the curves u, $u+du$, v, $v+dv$ is

$$dP = \frac{e^{-u}u^{l+m-1}\,du}{\Gamma(l+m)} \cdot \frac{v^{l-1}\,dv}{B(l, m)\,(1+v)^{l+m}}.$$

Since this is the probability that u and v will fall simultaneously in the intervals du and dv respectively, it follows that u is a $\gamma(l+m)$ variate, and v a $\beta_2(l, m)$ variate, and also that these variates are independent.

Example 1. Cauchy's distribution. The distribution of the quotient of two independent standard normal variates follows from the above theorem. For if z is this quotient, z^2 is the quotient of two independent Gamma variates, each of parameter $\frac{1}{2}$. Consequently z^2 is a $\beta_2(\frac{1}{2}, \frac{1}{2})$ variate, with probability differential

$$dP = \frac{(z^2)^{\frac{1}{2}-1}\,d(z^2)}{B(\frac{1}{2}, \frac{1}{2})\,(1+z^2)} = \frac{d(z^2)}{\pi(1+z^2)\sqrt{z^2}}.$$

The distribution of z follows immediately from this. For, since the range of z is from $-\infty$ to $+\infty$ while that of z^2 is from 0 to $+\infty$, the probability differential of z is

$$dp = \frac{dz}{\pi(1+z^2)},$$

the factor 2 disappearing, since the integral from $-\infty$ to $+\infty$ must be equal to unity. This distribution is associated with the name of Cauchy.

Example 2. Show that, if x and y are independent normal variates with means m_1, m_2 and variances σ_1^2, σ_2^2 respectively, the quotient

$$z = (x - m_1)/(y - m_2)$$

conforms to the distribution

$$dp = \frac{\sigma_1 \sigma_2\, dz}{\pi(\sigma_1^2 + \sigma_2^2 z^2)},$$

with a range $-\infty$ to $+\infty$.

COLLATERAL READING

COCHRAN, 1934, 6, pp. 178–91.
PITMAN, 1937, 8, pp. 216–18.
SAWKINS, 1940, 3, pp. 209–17.

EXAMPLES VIII

1. We shall see later (cf. § 82) that, if r is the coefficient of correlation between the variates in a random sample of n pairs of values from an uncorrelated bivariate normal population, then r^2 is a $\beta_1(\frac{1}{2}, \frac{1}{2}(n-2))$ variate. Hence show that the mean value of r^2 for such samples is $1/(n-1)$, and the s.e. of r therefore $1/\sqrt{(n-1)}$. Also show that the probability differential of r is

$$dp = \frac{(1 - r^2)^{\frac{1}{2}(n-4)}\, dr}{B(\frac{1}{2}, \frac{1}{2}(n-2))}.$$

2. *Quotient of independent Gamma variates.* Theorem VIII may be proved by direct integration, as in the cases of Theorems II, V and VI. Let $z = x/y$, where x and y are independent $\gamma(l)$ and $\gamma(m)$ variates. Then for a fixed value of y in the interval dy, $dx = y\,dz$, and the probability that z will lie in the interval dz is

$$dp = e^{-yz}(yz)^{l-1} y\, dz / \Gamma(l).$$

Consequently the probability that, for any value of y, z will lie in the interval dz is

$$dP = z^{l-1}\, dz \int_0^\infty y^{l+m-1} e^{-y(1+z)}\, dy / \Gamma(l)\, \Gamma(m)$$

$$= \frac{z^{l-1}\, dz}{B(l, m)\, (1+z)^{l+m}},$$

showing that z is a $\beta_2(l, m)$ variate as required.

3. Show that, for the $\gamma(l)$ distribution,

$$(\text{mean} - \text{mode})/\sigma = 1/\sqrt{l} = \tfrac{1}{2}\mu_3/\sigma^3.$$

Hence some writers prefer $\tfrac{1}{2}\mu_3/\sigma^3$ to μ_3/σ^3 as a definition of skewness. Show that the excess of kurtosis of the distribution is $6/l$.

4. Show that the mean value of the positive square root of a $\gamma(l)$ variate is $\Gamma(l+\tfrac{1}{2})/\Gamma(l)$. Hence prove that the mean deviation of a normal variate from its mean is $\sigma\sqrt{(2/\pi)}$.

5. Prove that the mean value of the positive square root of a $\beta_1(l, m)$ variate is $\Gamma(l+\tfrac{1}{2})\,\Gamma(l+m)/\Gamma(l)\,\Gamma(l+m+\tfrac{1}{2})$. Hence, using the distribution of r^2 given in Ex. 1, for samples from an uncorrelated bivariate normal population, show that the mean value of $|r|$ is $\Gamma(\tfrac{1}{2}(n-1))/\Gamma(\tfrac{1}{2}n)\sqrt{\pi}$.

6. A simple sample of n values is drawn from a population with a $\gamma(l)$ distribution. Show that, if $\bar{x}$ is the mean of the sample, $n\bar{x}$ is a $\gamma(nl)$ variate; and deduce that $E(\bar{x}) = l$, and that the sampling variance of $\bar{x}$ is l/n.

7. A simple sample of n values is drawn from a population with the exponential distribution whose probability density is ae^{-ax}, $(0 \leqslant x)$. Show that, if $\bar{x}$ is the mean of the sample, $na\bar{x}$ is a $\gamma(n)$ variate; and deduce that $E(\bar{x}) = 1/a$, and that the s.e. of $\bar{x}$ is $1/(a\sqrt{n})$.

8. Show that, if v is the square of a $\gamma(l)$ variate, its probability differential is $dp = v^{\frac{1}{2}(l-2)}e^{-\sqrt{v}}\,dv/2\Gamma(l)$.

9. Show that, if $x_i\,(i = 1, ..., n)$ are n independent Gamma variates with parameters m_i, any homogeneous function $F(x_1, ..., x_n)$ of these variates, of degree 0, is distributed independently of the sum $\sum_i x_i$. (Cf. Pitman, 1937, 8, pp. 216–17.)

This may be done by considering the m.g.f. of the simultaneous distribution of F and $\sum x_i$. As explained in §31 this m.g.f.

$$M(t_1, t_2) = E(\exp{(t_1 \sum x_i + t_2 F)})$$
$$= C \int_0^\infty \ldots \int_0^\infty \exp{(-\sum x_i)}\, x_1^{m_1-1} \ldots x_n^{m_n-1}$$
$$\times \exp{(t_1 \sum x_i + t_2 F)}\,dx_1 \ldots dx_n,$$

where $\quad 1/C = \Gamma(m_1) \ldots \Gamma(m_n).$

Substituting $x_i(1-t_1) = y_i$ we have

$$M(t_1, t_2) = C(1-t_1)^{-\Sigma m_i} \int_0^\infty \cdots \int_0^\infty \exp\left(-\sum y_i\right) y_1^{m_1-1} \cdots y_n^{m_n-1}$$
$$\times \exp\left(t_2 F(y_1, \ldots, y_n)\right) dy_1 \ldots dy_n,$$

since F is homogeneous of degree zero. Thus

$$M(t_1, t_2) = (\text{function of } t_1)\,(\text{function of } t_2),$$

and the variates F and $\sum x_i$ are therefore independent.

10. Given that the incomplete Beta function $B_x(l, m)$ is defined by

$$B_x(l, m) = \int_0^x x^{l-1}(1-x)^{m-1}\,dx,$$

and that $I_x(l, m) = B_x(l, m)/B(l, m)$, prove the relations

$$I_x(l, m) = 1 - I_{1-x}(m, l)$$

and $\qquad I_x(l+1, m+1) = x I_x(l, m+1) + (1-x)\, I_x(l+1, m).$

11. *Simultaneous sampling distribution of the mean and the variance.* The independence of the sampling distributions of the mean $\bar{x}$ and the variance S^2, of a random sample of n values from a normal population, was proved by Fisher from the simultaneous sampling distribution of these statistics. The probability that the n values of the sample will fall in the respective intervals $dx_1, \ldots, dx_n$ is, by the theorem of compound probability,

$$dp = (\sigma\sqrt{(2\pi)})^{-n} \exp\left[-\sum_i (x_i - \mu)^2/2\sigma^2\right] dx_1 dx_2 \ldots dx_n,$$

the n values being chosen independently from the normal population of mean μ and variance σ^2. But

$$\sum (x_i - \mu)^2 = \sum (x_i - \bar{x})^2 + n(\bar{x} - \mu)^2 = nS^2 + n(\bar{x} - \mu)^2,$$

so that

$$dp = (\sigma\sqrt{(2\pi)})^{-n} \exp\left[-n(\bar{x} - \mu)^2/2\sigma^2\right] \exp\left(-nS^2/2\sigma^2\right) dx_1 \ldots dx_n.$$

Fisher proved by geometrical reasoning* that this probability differential is expressible as

$$dp = C \exp[-n(\bar{x}-\mu)^2/2\sigma^2]\,d\bar{x}\exp(-nS^2/2\sigma^2)\,S^{n-2}\,dS,$$

where C is a constant. Since this is of the form

$$dp = \phi_1(\bar{x})\,d\bar{x}\,.\,\phi_2(S)\,dS,$$

the distributions of $\bar{x}$ and S are independent. The forms of $\phi_1(\bar{x})$ and $\phi_2(S)$ show that $\bar{x}$ is normally distributed, with mean μ and variance σ^2/n, and that $\frac{1}{2}nS^2/\sigma^2$ is a Gamma variate with parameter $\frac{1}{2}(n-1)$.

Another proof of the independence of $\bar{x}$ and S^2 will be given in §77.

12. Show that the rth moment of a $\beta_1(l,m)$ distribution about $x = 0$ is

$$\frac{l(l+1)\ldots(l+r-1)}{(l+m)(l+m+1)\ldots(l+m+r-1)}.$$

13. Show that, if r is less than m, the rth moment of a $\beta_2(l,m)$ distribution about $x = 0$ is

$$\frac{l(l+1)\ldots(l+r-1)}{(m-1)(m-2)\ldots(m-r)}.$$

14. Defining the harmonic mean (H.M.) of a variate x as the reciprocal of the expected value of $1/x$ show that, if $l > 1$, the H.M. of a $\gamma(l)$ variate is $l-1$, that of a $\beta_1(l,m)$ variate is $(l-1)/(l+m-1)$, and that of a $\beta_2(l,m)$ variate is $(l-1)/m$, m being positive.

---

* Fisher, 1925, 1, pp. 92–3.

CHAPTER IX

CHI-SQUARE AND SOME APPLICATIONS

74. Chi-square and its distribution

We have already seen (§ 68, Theorem IV) that if x_i ($i = 1, 2, ..., n$) are n independent variates normally distributed about a common mean zero with standard deviations σ_i, and

$$\chi^2 = \sum_i x_i^2/\sigma_i^2, \tag{1}$$

then $\frac{1}{2}\chi^2$ is a Gamma variate with parameter $\frac{1}{2}n$. The distribution of χ^2 is therefore

$$dP = \frac{1}{\Gamma(\frac{1}{2}n)} (\tfrac{1}{2}\chi^2)^{\frac{1}{2}(n-2)} \exp\left(-\tfrac{1}{2}\chi^2\right) d(\tfrac{1}{2}\chi^2), \tag{2}$$

which expresses the probability that the value of χ^2 found from a random sample will fall in the interval $d\chi^2$. This distribution is often referred to as the χ^2 distribution, and a variate conforming to it is said to be distributed like χ^2. Since χ^2 is twice a $\gamma(\frac{1}{2}n)$ variate, its mean value is n and its modal value $n-2$, as proved in § 67. Similarly, the variance of χ^2 is $2n$.

The distribution (2) was discovered by Helmert in 1875, and rediscovered independently in 1900 by Karl Pearson, who devised by means of it the χ^2 test of 'goodness of fit' which will soon be considered. We must first, however, examine the effect of one or more linear relations between the variates x_i; and, to make this clearer, we shall digress briefly to remind the reader of the properties of an orthogonal linear transformation.

75. Orthogonal linear transformation

Let the n variates x_i be subjected to the linear transformation

$$\xi_i = \sum_j c_{ij}x_j \quad (i, j = 1, ..., n). \tag{3}$$

If the constant coefficients c_{ij} are such that

$$\sum \xi_i^2 = \sum x_i^2 \tag{4}$$

the transformation is said to be orthogonal. In this case the coefficients satisfy the relations

$$\sum_i c_{ij}^2 = 1 = \sum_j c_{ij}^2 \tag{5}$$

and

$$\sum_i c_{ij} c_{ik} = 0 = \sum_i c_{ji} c_{ki} \quad (j \neq k). \tag{6}$$

These relations may be expressed verbally by saying that, in the determinant $|c_{ij}|$ of the coefficients, the sum of the squares of the elements in any row or in any column is equal to unity, while the sum of the products of corresponding elements in any two rows or in any two columns is equal to zero. The determinant of the coefficients is equal to ± 1; and consequently the Jacobian of the ξ's with respect to the x's is also ± 1. By changing the sign of one of the ξ's, if necessary, we may ensure that the value is $+1$; and we shall assume in all cases that this has been done. It follows that

$$\frac{\partial(x_1, \ldots, x_n)}{\partial(\xi_1, \ldots, \xi_n)} = 1. \tag{7}$$

It is easy to prove that, if the x's are statistically independent variates, normally distributed about zero with unit s.d., so are the ξ's. For, the probability that simultaneously the n values of the variates x_i will fall in the respective intervals dx_i is the product of the probabilities for the individual variates, and is therefore given by

$$dP = (2\pi)^{-\frac{1}{2}n} \exp\left(-\tfrac{1}{2}\sum_i x_i^2\right) dx_1 dx_2 \ldots dx_n.$$

The probability density for the joint distribution of the variates x_i is therefore

$$(2\pi)^{-\frac{1}{2}n} \exp\left(-\tfrac{1}{2}\sum_i x_i^2\right) = (2\pi)^{-\frac{1}{2}n} \exp\left(-\tfrac{1}{2}\sum_i \xi_i^2\right).$$

But the 'volume' of the element bounded by the n pairs of hypersurfaces ξ_i, $\xi_i + d\xi_i$ is

$$\frac{\partial(x_1, \ldots, x_n)}{\partial(\xi_1, \ldots, \xi_n)} d\xi_1 d\xi_2 \ldots d\xi_n = d\xi_1 d\xi_2 \ldots d\xi_n,$$

so that the probability that the ξ's will fall simultaneously in their respective intervals $d\xi_i$ is expressible as

$$dP = (2\pi)^{-\frac{1}{2}} \exp\left(-\tfrac{1}{2}\xi_1^2\right) d\xi_1 \dots (2\pi)^{-\frac{1}{2}} \exp\left(-\tfrac{1}{2}\xi_n^2\right) d\xi_n.$$

From the form of this expression it follows that the ξ's are statistically independent, and are normally distributed about zero with unit s.d.

The transformation (3) corresponds to a rotation of rectangular axes in Euclidean space of n dimensions. In the choice of the coefficients c_{ij} there is a degree of arbitrariness. For instance the coefficients for ξ_1 may be chosen first, subject only to the condition

$$\sum_i c_{1i}^2 = 1,$$

which ensures that the constants c_{1i} are the components of a unit vector. When ξ_1 has been determined ξ_2 may be chosen in an infinity of ways, subject only to the conditions

$$\sum_i c_{2i}^2 = 1, \quad \sum_i c_{1i} c_{2i} = 0,$$

which express that c_{2i} are components of a unit vector, orthogonal to the vector whose components are c_{1i}. At each step there is freedom of choice of an axis orthogonal to those already chosen; and only in the case of the nth axis is there no freedom of choice.

76. Linear constraints. Degrees of freedom

As above let the variates x_i be normally distributed about zero as mean, with unit s.d. To begin with we assume that they are functionally independent; but presently we shall impose on them the linear restriction

$$a_1 x_1 + a_2 x_2 + \dots + a_n x_n = 0, \tag{8}$$

an equation which may be divided throughout by the constant necessary to make $\sum_i a_i^2 = 1$; and we assume that this has been done.

If the variates x_i are independent, we may obtain by an orthogonal linear transformation a set of n statistically independent variates ξ_i, each of which is normally distributed about zero as mean, with unit s.d.; and this may be done so that ξ_1 is identical with the first member of (8). Now let the condition (8) be imposed on the x's. This is equivalent to putting $\xi_1 = 0$; and, since the variates $\xi_2, \ldots, \xi_n$ are statistically independent of ξ_1, their distributions are unaltered by the condition so imposed. Consequently

$$\chi^2 = \sum_i x_i^2 = \sum_1^n \xi_i^2 = \sum_2^n \xi_i^2 = \sum_{i=2}^n 2u_i,$$

where the u_i are Gamma variates, each with parameter $\frac{1}{2}$. Thus $\frac{1}{2}\chi^2$ is the sum of $n-1$ independent $\gamma(\frac{1}{2})$ variates, and is therefore itself a Gamma variate with parameter $\frac{1}{2}(n-1)$. The distribution of χ^2 is therefore obtained from (2) by putting $n-1$ in place of n. The condition (8) has reduced the number of independent variates by one. The number of independent variates is usually called the number of *degrees of freedom* (D.F.), or briefly the number of *freedoms*. The term is borrowed from geometry and mechanics, where the position of a point or of a body is specified by a number of functionally independent variables called coordinates. Each independent co-ordinate corresponds to one degree of freedom of movement. Any constraint on the body reduces the number of degrees of freedom. For this reason a linear relation between the variates x_i is called a *linear constraint*. We shall assume that only linear constraints are involved.

An appeal to geometry throws light on the above reasoning. If the variables x_i are regarded as rectangular Cartesian coordinates of the current point P in Euclidean space of n dimensions, the square of the distance of P from the origin O is

$$OP^2 = \sum x_i^2 = \chi^2,$$

and, in terms of the alternative set ξ_i of coordinates relative to rectangular axes through the same origin,

$$OP^2 = \sum \xi_i^2 = \chi^2.$$

When the variables are connected by the equation (8), the point P is constrained to lie on a hyperplane through the origin, determined by this equation. In terms of the ξ's the equation of the hyperplane is simply

$$\xi_1 = 0,$$

and for a point P on this hyperplane we have

$$\chi^2 = OP^2 = \sum_1^n \xi_i^2 = \sum_2^n \xi_i^2$$

as above. Thus the linear constraint (8) restricts the freedom of movement of P to the hyperplane, in which there are only $n-1$ independent coordinates. χ^2 is then said to correspond to $n-1$ D.F.

That each linear constraint reduces the number of freedoms by unity may be shown in a similar manner. Thus if there is a second linear constraint

$$b_1' x_1 + b_2' x_2 + \dots + b_n' x_n = 0, \tag{9}$$

it is expressible in terms of the ξ's in the form

$$b_2 \xi_2 + b_3 \xi_3 + \dots + b_n \xi_n = 0, \tag{10}$$

in which $\sum_2^n b_i^2 = 1$. We obtain χ^2 as the sum of squares of $n-1$ standard normal variates ξ_i $(i = 2, \dots, n)$, which are now connected by the linear constraint (10). Proceeding as before by an appropriate orthogonal transformation

$$\eta_i = \sum_{j=2}^n c_{ij} \xi_j,$$

in which η_2 is identical with the first member of (10), we may express χ^2 as

$$\chi^2 = \sum_{i=3}^n \eta_i^2,$$

and $\frac{1}{2}\chi^2$ is thus a Gamma variate with parameter $\frac{1}{2}(n-2)$, so that χ^2 corresponds to $n-2$ D.F. The argument holds for any number of linear constraints. Following Yule and Kendall* we shall denote the number of freedoms by ν. Then, if the n variates are subject to m linear constraints,

$$\nu = n - m. \tag{11}$$

In place of (2) we thus have for the distribution of χ^2 corresponding to ν D.F.,

$$dP = \frac{1}{2^{\frac{1}{2}\nu} \Gamma(\frac{1}{2}\nu)} (\chi^2)^{\frac{1}{2}(\nu-2)} \exp\left(-\frac{1}{2}\chi^2\right) d\chi^2. \tag{12}$$

* 1937, 1, p. 415.

Since $\frac{1}{2}\chi^2$ is a $\gamma(\frac{1}{2}\nu)$ variate, the mean value of χ^2 is ν and its modal value is $\nu - 2$. The probability that the value of χ^2 from a random sample will not exceed a fixed value χ_0^2 is obtained by integrating (12) with respect to χ^2 from 0 to χ_0^2. Similarly, the probability that χ^2 will exceed χ_0^2 is the integral* of (12) with respect to χ^2 from χ_0^2 to ∞. The accompanying Table 3 gives the values of χ_0^2, for values of ν from 1 to 30, and for various fixed values of the probability P of exceeding χ_0^2. In other words, P is the probability that in random sampling the value of χ^2 shown in the body of the table will be exceeded.

Example. Show that, for 2 D.F., the probability P of a value of χ^2 greater than χ_0^2 is $\exp(-\frac{1}{2}\chi_0^2)$, and hence that $\chi_0^2 = 2\log_e 1/P$.

77. Distribution of the 'sum of squares' for a random sample from a normal population

Consider a random sample of n independent values x_i from a normal population of variance σ^2. If as usual $\bar{x}$ denotes the mean of the sample, the sum of squares of the deviations from the mean is

$$nS^2 = \sum_i (x_i - \bar{x})^2. \tag{13}$$

This is the 'sum of squares' to be considered; and we shall prove that nS^2/σ^2 is distributed like χ^2 for $n-1$ D.F. We know that

$$\sum_i x_i^2 = \sum_i (x_i - \bar{x})^2 + n\bar{x}^2 = nS^2 + n\bar{x}^2. \tag{14}$$

If then we introduce an orthogonal linear transformation (3) of the variables x_i such† that $\xi_1 = \bar{x}\sqrt{n}$, we have by (14)

$$nS^2 = \sum_i x_i^2 - n\bar{x}^2 = \sum_{i=1}^n \xi_i^2 - \xi_1^2 = \sum_{i=2}^n \xi_i^2,$$

and therefore
$$nS^2/\sigma^2 = \sum_{i=2}^n \xi_i^2/\sigma^2. \tag{15}$$

Now, with origin at the mean of the population, the x's are normally distributed about zero as mean with S.D. σ, and so also are the ξ's.

* For a method evaluating this integral see Fisher, 1935, 1, pp. 356–7. See also Ex. IX, 9.

† Cf. Sawkins, 1940, 3, p. 225.

Consequently $\frac{1}{2}nS^2/\sigma^2$ is a Gamma variate with parameter $\frac{1}{2}(n-1)$, and its distribution is

$$dP = \frac{1}{\Gamma(\frac{1}{2}(n-1))} \left(\frac{nS^2}{2\sigma^2}\right)^{\frac{1}{2}(n-3)} \exp\left(\frac{-nS^2}{2\sigma^2}\right) d\left(\frac{nS^2}{2\sigma^2}\right).$$

In other words nS^2/σ^2 is distributed like χ^2 with $n-1$ D.F.

It is convenient to express the result in terms of the statistic s^2 of § 54, which is an unbiased estimate of the variance of the population. Thus

$$nS^2 = (n-1)s^2 = \nu s^2,$$

and the distribution of s^2 is therefore

$$dP = \frac{1}{\Gamma(\frac{1}{2}\nu)} \left(\frac{\nu}{2\sigma^2}\right)^{\frac{1}{2}\nu} (s^2)^{\frac{1}{2}(\nu-2)} \exp\left(-\frac{\nu s^2}{2\sigma^2}\right) ds^2, \qquad (16)$$

the estimate s^2 being based on ν D.F.

The coefficient of ds^2 in (16) is the probability density in the distribution of s^2. Suppose that the population variance is unknown, and that we enquire what value of it would make the probability density of s^2 a maximum for the given value of s^2. This is obtained by equating to zero its derivative with respect to σ^2. The procedure leads to $\sigma^2 = s^2$. For this reason s^2 is called the *optimum value* of the population variance corresponding to the given sample; and the method of obtaining it is called the method of *maximum likelihood*. It is due to R. A. Fisher.

In the above argument ξ_1 is independent of $\xi_2, ..., \xi_n$ and therefore, in virtue of (15), $\bar{x}$ is independent of S^2. Thus the sampling distributions of the mean and the variance are independent. And, since ξ_1/σ is a standard normal variate, so is $\bar{x}\sqrt{n}/\sigma$. It follows that $\bar{x}$ is normally distributed with the same mean as the population, and with variance σ^2/n.

78. Nature of the chi-square test. An illustration

The χ^2 test is a means of judging the credibility of an hypothesis concerning the population (or populations) from which the values of the sample (or samples) are drawn. The hypothesis to be tested must be of such a nature that we can determine, from the sample values of the variate, the corresponding value of a certain statistic, which is distributed like χ^2 for a known number of freedoms. The

Chi-Square Table

TABLE 3. *Values of χ^2 with probability P of being exceeded in random sampling*

ν = number of degrees of freedom

P ν	0·99	0·95	0·50	0·30	0·20	0·10	0·05	0·01
1	0·0002	0·004	0·46	1·07	1·64	2·71	3·84	6·64
2	0·020	0·103	1·39	2·41	3·22	4·60	5·99	9·21
3	0·115	0·35	2·37	3·66	4·64	6·25	7·82	11·34
4	0·30	0·71	3·36	4·88	5·99	7·78	9·49	13·28
5	0·55	1·14	4·35	6·06	7·29	9·24	11·07	15·09
6	0·87	1·64	5·35	7·23	8·56	10·64	12·59	16·81
7	1·24	2·17	6·35	8·38	9·80	12·02	14·07	18·48
8	1·65	2·73	7·34	9·52	11·03	13·36	15·51	20·09
9	2·09	3·32	8·34	10·66	12·24	14·68	16·92	21·67
10	2·56	3·94	9·34	11·78	13·44	15·99	18·31	23·21
11	3·05	4·58	10·34	12·90	14·63	17·28	19·68	24·72
12	3·57	5·23	11·34	14·01	15·81	18·55	21·03	26·22
13	4·11	5·89	12·34	15·12	16·98	19·81	22·36	27·69
14	4·66	6·57	13·34	16·22	18·15	21·06	23·68	29·14
15	5·23	7·26	14·34	17·32	19·31	22·31	25·00	30·58
16	5·81	7·96	15·34	18·42	20·46	23·54	26·30	32·00
17	6·41	8·67	16·34	19·51	21·62	24·77	27·59	33·41
18	7·02	9·39	17·34	20·60	22·76	25·99	28·87	34·80
19	7·63	10·12	18·34	21·69	23·90	27·20	30·14	36·19
20	8·26	10·85	19·34	22·78	25·04	28·41	31·41	37·57
21	8·90	11·59	20·34	23·86	26·17	29·62	32·67	38·93
22	9·54	12·34	21·34	24·94	27·30	30·81	33·92	40·29
23	10·20	13·09	22·34	26·02	28·43	32·01	35·17	41·64
24	10·86	13·85	23·34	27·10	29·55	33·20	36·42	42·98
25	11·52	14·61	24·34	28·17	30·68	34·38	37·65	44·31
26	12·20	15·38	25·34	29·25	31·80	35·56	38·88	45·64
27	12·88	16·15	26·34	30·32	32·91	36·74	40·11	46·96
28	13·56	16·93	27·34	31·39	34·03	37·92	41·34	48·28
29	14·26	17·71	28·34	32·46	35·14	39·09	42·56	49·59
30	14·95	18·49	29·34	33·53	36·25	40·26	43·77	50·89

Reproduced by permission of the author, Professor R. A. Fisher, from his book on *Statistical Methods for Research Workers.*

table then tells us the probability P that, in random sampling, a value of this statistic will occur greater than the value actually obtained. If this probability is very small, we regard the value obtained as significantly large, and conclude that the hypothesis is probably incorrect.

Two conventional values of P are employed in deciding significance, viz. 0·05 and 0·01. These determine the 5 % and the 1 % levels of significance respectively. If the value of P obtained is greater than 0·05 we infer that, in more than 5 % of random samples from the population in question, the value of χ^2 obtained would be greater than that actually found, which is therefore regarded as not significantly large. If, however, P is less than 0·05 the value found is regarded as significant at that level. Similar remarks apply to the 1 % level of significance. Values which are significant at the 1 % level of probability are said to be highly significant, and are sometimes distinguished by a double asterisk. Significance at the 5 % level is then denoted by a single asterisk.

The value of χ^2 obtained from the sample may also be significantly small. A glance at the probability curve of χ^2 will help to make this point clear. When the number ν of degrees of freedom is

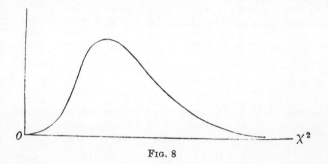

FIG. 8

greater than 2, this curve has the form indicated in the diagram, except that when $\nu = 3$ it touches the y-axis at the origin, and when $\nu = 4$ it touches neither axis at that point. The ordinate for any value of χ^2 is the probability density for that value; and this is small for small values of χ^2. When the value of P found from the table is greater than 0·95, the probability of obtaining a smaller value of χ^2 is less than 5 %, and the sample value must be regarded as significantly small. Similarly, when P is greater than 0·99 the probability of a smaller value of χ^2 is less than 1 %, and the smallness of the sample value is highly significant.

As an illustration of the χ^2 test we may consider the following. Let x_i be a random sample of n values from a normal population of variance σ^2, $\bar{x}$ the sample mean and

$$nS^2 = \sum (x_i - \bar{x})^2.$$

Then we know that nS^2/σ^2 is distributed like χ^2 with $n-1$ D.F. If then we are given the sample values x_i, but have no certain knowledge of the population from which they were drawn, we may test the hypothesis that they came from a normal population of S.D. σ. For, when S^2 has been found from the sample, our hypothesis gives nS^2/σ^2 as the sample value of χ^2; and the table then enables us to decide whether this value is significant or not, that is to say whether our hypothesis is improbable or not.

Example. A random sample of 12 values gave an unbiased estimate s^2 of the population variance equal to $10 \cdot 62$ mm.^{2} May the sample be reasonably regarded as from a normal population with variance 7 mm.^{2}?

Here $nS^2 = 11 \times 10 \cdot 62 = 116 \cdot 82$, and, according to our hypothesis concerning the population,

$$\chi^2 = 116 \cdot 82/7 = 16 \cdot 7.$$

The number of D.F. is 11. From the table we see that the probability, P, that the value of χ^2 in such samples will exceed $16 \cdot 7$ is greater than $0 \cdot 10$. The value found is therefore not significant, and the test provides no evidence against the hypothesis of a normal population with variance 7.

79. Test of goodness of fit

We shall now consider the χ^2 test of goodness of fit devised by Karl Pearson. As in Chapter VII let the population be one whose members may be separated into a number k of classes, and let p_i be the relative frequency of the ith class. In the choice of a simple sample of n members from this population, the probability at any drawing that the individual selected will belong to the ith class is p_i. The frequency f_i of this class in sampling has a mean value m_i given by

$$m_i = np_i, \tag{17}$$

and the distribution of f_i is the binomial. For large values of n the binomial distribution approximates to normal; and the sampling distribution of any class frequency is therefore approximately normal for large samples.

Next suppose that we are given a set of class frequencies f_i, whose sum is n, without any information about the source of the values. We may wish to enquire whether they may reasonably be regarded as those of a simple sample from a certain hypothetical population. The population is hypothetical in the sense that it is determined by an hypothesis, which must be such as to enable us to calculate the expected values m_i of the class frequencies in random sampling from it. We shall prove that, for large samples, the variate $\sum_i (f_i - m_i)^2/m_i$ conforms approximately to the χ^2 distribution, and shall show how the number of degrees of freedom is determined. The value of χ^2 obtained from the sample enables us to test the credibility of our hypothesis.

Since, for large samples, the class frequency f_i is distributed approximately normally about m_i as mean, the variate

$$x_i = f_i - m_i \qquad (18)$$

is distributed approximately normally about zero as mean. We require the variance σ_i^2 of f_i when the class frequencies are entirely *independent*; and, to obtain this, we must take account* of the variation in their sum n. Thus

$$n = \sum f_i, \qquad (19)$$

and the size n of the sample varies about a mean $\bar{n}$ with s.d. σ_n. Since the frequencies f_i are supposed independent, the variance of n is given by

$$\sigma_n^2 = \sum_i \sigma_i^2. \qquad (20)$$

Now, in virtue of § 47 (8), the variance σ_i^2 for samples of varying size is

$$\sigma_i^2 = \bar{n} p_i q_i + p_i^2 \sigma_n^2. \qquad (21)$$

Summing for all the classes we have, by (20),

$$\sigma_n^2 = \bar{n} \sum_i p_i q_i + \sigma_n^2 \sum_i p_i^2,$$

or, since $\sum p_i = 1$,

$$\sigma_n^2 \sum (p_i - p_i^2) = \bar{n} \sum p_i q_i.$$

Hence $\sigma_n^2 = \bar{n}$, so that (21) is equivalent to

$$\sigma_i^2 = (p_i q_i + p_i^2) \bar{n} = p_i \bar{n} = m_i. \qquad (22)$$

The quantity defined by

$$\chi^2 = \sum x_i^2/\sigma_i^2 = \sum (f_i - m_i)^2/m_i \qquad (23)$$

is a sum of squares of standard approximately normal deviates, and is therefore distributed approximately like χ^2.

* Cf. Fisher, 1922, 1, p. 88.

For samples of fixed size, the number of D.F. is clearly less than the number k of classes. For the sum of the class frequencies is constant, and this corresponds to a linear constraint on the variates x_i. Further, to determine the theoretical class frequencies m_i, it is sometimes necessary to estimate parameters of the population from the data of the sample. For instance, in testing the hypothesis of a normal population, it may be necessary to estimate the mean and the variance of the population from the sample values. Each estimate of a parameter obtained in this manner corresponds to the introduction of a linear constraint. For the moments about the mean are linear, or approximately linear, functions of the class frequencies; and, in equating such a function to a parameter, we introduce an approximately linear constraint on the variates x_i. In calculating the number of freedoms of χ^2, each constraint introduced in this manner must be recognized.

The approximation of the binomial distribution to normal, when n is large, does not hold for very small values of p or q. Hence the above argument is not valid if one of the class frequencies is small; for that would make the corresponding relative frequency f/n very small, and therefore the probability p for that class in sampling from the population also very small. Classes of small frequency may be treated by combining two or more of them to form a class sufficiently large.

80. Numerical examples

Example 1. Can the wages of 1,000 employees, given in Ex. I, 3 (p. 17), be regarded as a random sample from a normal population?

Our hypothesis is that the population is normal. Since the sample is large its mean and its variance are taken as estimates of those of the population. In Ex. III, 6 (p. 60) the class frequencies per thousand of the normal population are given. On account of the smallness of the extreme frequencies we combine the first two classes, and also the last two, leaving 13 classes. Since the sum of the class frequencies is constant, and two of the parameters were estimated from the sample, the number of degrees of freedom is 10. The value of χ^2 from the sample is

$$\chi^2 = \frac{5^2}{18} + \frac{10^2}{25} + 0 + \frac{14^2}{79} + \dots + \frac{(3\cdot2)^2}{18} = 19\cdot84.$$

From the table we see that, for 10 D.F., this value of χ^2 is significant at the 5 % level. We conclude that the assumption of a normal population is probably incorrect.

Example 2. From the adult male populations of seven large cities, random samples of the sizes indicated below were taken, and the numbers of married and single men recorded. Do the data indicate any significant variation among the cities in the tendency of men to marry?

City...	A	B	C	D	E	F	G	Total
Married	133	164	155	106	153	123	146	980
Single	36	57	40	37	55	33	36	294
Total	169	221	195	143	208	156	182	1274

We test the hypothesis that there is no significant variation in the tendency mentioned. Then the men from each city may be regarded as a simple sample from a population in which the ratio of married men to single is approximately the same as in the column of totals. This ratio is $10:3$. The theoretical frequencies for any city are then obtained by dividing the total for that city into two parts in this ratio. These frequencies are:

City...	A	B	C	D	E	F	G	Total
Married	130	170	150	110	160	120	140	980
Single	39	51	45	33	48	36	42	294
Total	169	221	195	143	208	156	182	1274

From these figures we have

$$\chi^2 = \frac{3^2}{130} + \frac{6^2}{170} + \frac{5^2}{150} + \ldots + \frac{6^2}{42} = 5\cdot34.$$

To find the number of freedoms we observe that the sum of the frequencies of married and single men from any city is constant, being equal to the size of the sample from that city. This reduces the number of independent frequencies to 7. And further, a parameter of the population was estimated from the sample, namely, the ratio of the numbers of married and single men. Consequently $\nu = 6$. For this number of freedoms the probability of obtaining a larger value of χ^2 than $5\cdot34$ is about $0\cdot50$. The value is therefore not significant, and the test furnishes no evidence against the hypothesis.

Example 3. May the data in Ex. III, 7 (p. 61) be regarded as those of a random sample from a Poissonian distribution?

The mean, $m = 1\cdot2$, was estimated from the sample, and from it the theoretical frequencies per thousand of the Poissonian distribution were calculated in the example referred to. To apply the χ^2 test we combine the last three classes. Then

$$\chi^2 = \frac{(3\cdot8)^2}{301\cdot2} + \frac{(3\cdot6)^2}{361\cdot4} + \ldots + \frac{(4\cdot4)^2}{7\cdot6} = 3\cdot5.$$

Here the number of classes is 6; but the total frequency is constant, and m was estimated from the sample, so that $\nu = 4$. With 4 D.F. the probability that χ^2 will exceed $3\cdot5$ is nearly $0\cdot50$. The value is therefore not at all significant, and the assumption of a Poissonian population is not discredited.

81. Additive property of chi-square

THEOREM I. *If the independent variates x and y conform to the χ^2 distribution, with ν_1 and ν_2 D.F. respectively, then $x+y$ is distributed like χ^2 with $\nu_1 + \nu_2$ D.F.*

For $\frac{1}{2}x$ and $\frac{1}{2}y$ are independent Gamma variates with parameters $\frac{1}{2}\nu_1$ and $\frac{1}{2}\nu_2$ respectively. Therefore, by § 68, Theorem II, $\frac{1}{2}(x+y)$ is a Gamma variate with parameter $\frac{1}{2}(\nu_1 + \nu_2)$. Consequently $x+y$ conforms to the χ^2 distribution with $\nu_1 + \nu_2$ D.F.

Similarly, corresponding to § 68, Theorem III, we may state

THEOREM II. *If the sum of two independent positive variates is distributed like χ^2 with $\nu_1 + \nu_2$ D.F., and one of them is distributed like χ^2 with ν_1 D.F., then the other is distributed like χ^2 with ν_2 D.F.*

In the same way Theorem V of § 69 and Theorem VIII of § 73 may be expressed as

THEOREM III. *If the independent variates x and y are distributed like χ^2 with ν_1 and ν_2 D.F. respectively, then $x/(x+y)$ is a Beta variate of the first kind with parameters $\frac{1}{2}\nu_1$ and $\frac{1}{2}\nu_2$, while x/y is a Beta variate of the second kind with the same parameters.*

82. Samples from an uncorrelated bivariate normal population. Distribution of the correlation coefficient

The distribution of the coefficient of correlation, r, in samples from a normally correlated bivariate population was given by Fisher[*] in 1915. In the particular case of an uncorrelated population ($\rho = 0$) the distribution of r is very simple. Let σ_1 and σ_2 be the S.D.'s of x and y in the population, and let the variates be measured from their means. Consider a random sample of n pairs of values x_t, y_t, from such a population. The variances S_1^2, S_2^2 of x, y in the sample are given by

$$nS_1^2 = \sum_t (x_t - \bar{x})^2, \quad nS_2^2 = \sum_t (y_t - \bar{y})^2,$$

and the correlation, r, in the sample is

$$r = \frac{\sum (x_t - \bar{x})(y_t - \bar{y})}{nS_1 S_2}.$$

---

[*] Fisher, 1915, 2. A simple and excellent proof of a different character has recently been published by Sawkins (1944, 1). The proof for $\rho = 0$ given in this section is based on that of Sawkins.

Owing to the nature of the sampling the n values y_t are independent. Let these be subjected to an orthogonal transformation yielding n variates η_t, the first of which may be taken as

$$\eta_1 = \frac{1}{\sqrt{n}} \sum_t y_t = \sqrt{n}\,\bar{y},$$

since the sum of squares of the coefficients of the y_t is unity. Then

$$\sum_1^n \eta_t^2 = \sum_1^n y_t^2 = \sum_1^n (y_t - \bar{y})^2 + n\bar{y}^2 = nS_2^2 + \eta_1^2,$$

so that
$$nS_2^2 = \sum_2^n \eta_t^2, \tag{i}$$

and this sum, divided by σ_2^2, is distributed like χ^2 with $n-1$ D.F. Further, from the above definition of r,

$$\sqrt{n}\,rS_2 = \sum (x_t - \bar{x})(y_t - \bar{y})/\sqrt{n}\,S_1 = \sum (x_t - \bar{x})\,y_t/\sqrt{n}\,S_1.$$

We may take this sum for the second variate, η_2; for the sum of the squares of the coefficients of the y_t is unity, the orthogonal condition is satisfied by the coefficients of η_1 and η_2, and the variables x and y are independent. Since $\eta_2^2 = nr^2 S_2^2$ it follows from (i) that $\sum_3^n \eta_t^2 = n(1 - r^2) S_2^2$. Further, the η_t/σ_2 are independent standard normal variates. Thus $nr^2 S_2^2/\sigma_2^2$ and $n(1 - r^2) S_2^2/\sigma_2^2$ are distributed independently like χ^2 with 1 and $n-2$ D.F. respectively; or we may express it by saying that the former is a χ_1^2 and the latter a χ_{n-2}^2. It follows that

$$r^2 = \frac{nr^2 S_2^2}{nS_2^2} = \frac{nr^2 S_2^2/\sigma_2^2}{nr^2 S_2^2/\sigma_2^2 + n(1 - r^2) S_2^2/\sigma_2^2} = \frac{\chi_1^2}{\chi_1^2 + \chi_{n-2}^2},$$

and therefore, by § 81, Theorem III, r^2 is a $\beta_1(\frac{1}{2}, \frac{1}{2}(n-2))$ variate, with distribution
$$dp = \frac{(r^2)^{\frac{1}{2}-1}(1 - r^2)^{\frac{1}{2}(n-4)}\,d(r^2)}{B(\frac{1}{2}, \frac{1}{2}(n-2))}. \tag{24}$$

We may thus state the important

THEOREM IV. *For random samples of n pairs of values from an uncorrelated bivariate normal population, r^2 is a Beta variate of the first kind with parameters $\frac{1}{2}$ and $\frac{1}{2}(n-2)$.*

The expected value of r^2 for such samples is therefore $1/(n-1)$, and the s.e. of r is $1/\sqrt{(n-1)}$. And since r ranges from -1 to 1, the opposite values $\pm r$ giving the same value of r^2, the distribution of r is

$$dp = \frac{(1-r^2)^{\frac{1}{2}(n-4)}\,dr}{B(\frac{1}{2},\frac{1}{2}(n-2))}. \qquad (25)$$

In considering the linear regression of y on x we proved the relation (cf. § 27 (31))

$$\Sigma\,(y_t-\bar{y})^2 = \Sigma\,(y_t-Y_t)^2 + \Sigma\,(Y_t-\bar{y})^2, \qquad (26)$$

where Y is the estimate of y from the regression equation. In the present notation this corresponds to the identity

$$nS_2^2 = n(1-r^2)\,S_2^2 + nr^2S_2^2. \qquad (27)$$

We have just seen that, for samples from the above population, these three sums are independent. Thus the sum of squares of deviations from the line of regression is distributed independently of the sum of squares of deviations due to regression in the sample. When divided by σ_2^2 the three sums in (27) are distributed like χ^2 with $n-1$, $n-2$ and 1 D.F. respectively, illustrating the additive property of χ^2 stated in § 81.

83. Distribution of regression coefficients and correlation ratios

The distribution of the linear *regression coefficient*, b, of y on x follows from the above results. For

$$b = rS_2/S_1$$

so that
$$\frac{b^2\sigma_1^2}{\sigma_2^2} = \frac{nr^2S_2^2/\sigma_2^2}{nS_1^2/\sigma_1^2}.$$

As we have just seen, the numerator and denominator of the second member are distributed independently like χ^2 with 1 and $n-1$ D.F. respectively. Consequently, by Theorem III, the quotient is a $\beta_2(\frac{1}{2},\frac{1}{2}(n-1))$ variate, and therefore $b^2\sigma_1^2/\sigma_2^2$ is a variate of the same type. Hence we may state

THEOREM V. *In random samples of n pairs of values from an uncorrelated bivariate normal population, in which the standard deviations of the variables are σ_1 and σ_2, the linear regression coefficient*

b of y on x is such that $b^2\sigma_1^2/\sigma_2^2$ is a *Beta variate of the second kind with parameters* $\frac{1}{2}$ *and* $\frac{1}{2}(n-1)$.

Since the mean value of a $\beta_2(l, m)$ variate is $l/(m-1)$, the expected value of b^2 in sampling is $\sigma_2^2/(n-3)\sigma_1^2$. And, since the range of b is from $-\infty$ to $+\infty$, the distribution of b is

$$dp = \frac{\sigma_1\sigma_2^{n-1}db}{B(\frac{1}{2}, \frac{1}{2}(n-1))(\sigma_2^2+b^2\sigma_1^2)^{\frac{1}{2}n}}. \tag{28}$$

The distribution of the *correlation ratio*, η, of y on x may be determined from the corresponding resolution of the sum of squares. With the notation of § 34 let the subscript i distinguish the particular array of y's, while j indicates position in that array, $\bar{y}_i$ denoting the mean of the y's in the ith vertical array, and n_i the frequency in that array. Then, in virtue of § 34, we have the resolution

$$\sum_i \sum_j (y_{ij}-\bar{y})^2 = \sum_i \sum_j (y_{ij}-\bar{y}_i)^2 + \sum_i n_i(\bar{y}_i-\bar{y})^2, \tag{29}$$

the various sums being equal to the corresponding terms of the identity

$$nS_2^2 = n(1-\eta^2)S_2^2 + n\eta^2 S_2^2. \tag{30}$$

Now the first member of (29), divided by σ_2^2, is distributed like χ^2 with $n-1$ D.F. Further, since $\sum_j$ denotes summation over the values of any particular array, $\sum_j (y_{ij}-\bar{y}_i)^2/\sigma_2^2$ is distributed like χ^2 with n_i-1 D.F. Summing for all the h arrays we see that the first sum in the second member of (29), divided by σ_2^2, is a χ^2 with $\sum_i (n_i-1)$ or $n-h$ D.F. And, because the means of the arrays are distributed independently of their variances, it follows from § 81, Theorem II, that the last sum in (29), divided by σ_2^2, is a χ^2 with $h-1$ D.F.* We may therefore write

$$\eta^2 = \frac{n\eta^2 S_2^2}{nS_2^2} = \frac{n\eta^2 S_2^2/\sigma_2^2}{n\eta^2 S_2^2/\sigma_2^2 + n(1-\eta^2)S_2^2/\sigma_2^2} = \frac{\chi_{h-1}^2}{\chi_{h-1}^2+\chi_{n-h}^2},$$

* See also Ex. 7 at the end of this chapter.

in which the subscript indicates the number of D.F. for the χ^2. Thus, in virtue of § 81, Theorem III, η^2 is a $\beta_1(\frac{1}{2}(h-1), \frac{1}{2}(n-h))$ variate, and we may state

THEOREM VI. *For random samples of n pairs of values from an uncorrelated bivariate normal population, the square of the correlation ratio of y on x is a Beta variate of the first kind with parameters $\frac{1}{2}(h-1)$ and $\frac{1}{2}(n-h)$, where h is the number of arrays of y's.*

The distribution of η^2 is thus

$$dp = \frac{(\eta^2)^{\frac{1}{2}(h-3)}(1-\eta^2)^{\frac{1}{2}(n-h-2)}d(\eta^2)}{B(\frac{1}{2}(h-1), \frac{1}{2}(n-h))}, \tag{31}$$

and its mean value,* by § 69 (23), is

$$E(\eta^2) = \frac{h-1}{n-1}. \tag{32}$$

COLLATERAL READING

YULE and KENDALL, 1937, 1, chapter XXII.
MILLS, 1938, 1, pp. 618–36.
FISHER, 1938, 2, chapter IV.
AITKEN, 1939, 1, pp. 99–105.
RIDER, 1939, 6, chapter VII.
KENNEY, 1939, 3, part II, pp. 164–71.
TIPPETT, 1931, 2, chapter IV.
KENDALL, 1943, 2, chapter XII.
SNEDECOR, 1938, 3, chapter IX.
GOULDEN, 1939, 2, chapters IX and X.
FISHER, 1922, 1 and 1935, 1, pp. 353–4, and 356–8.
PEARSON, K., 1900, 1 and 1922, 3.
SAWKINS, 1940, 3, 1941, 1 and 1944, 1.

EXAMPLES IX

1. Apply the χ^2 test to show that the deaths of centenarians recorded in the example of § 18 may reasonably be regarded as a random sample from a Poissonian population.

2. A certain hypothesis was tested by two similar experiments, which gave $\chi^2 = 14\cdot7$ for $\nu = 9$ and $\chi^2 = 14\cdot9$ for $\nu = 11$. Show that the two experiments combined give less reason for confidence in the hypothesis than either experiment alone.

* Cf. Fisher, 1922, 2, pp. 604–5.

3. Prove by mathematical induction that the sum of the squares of n independent standard normal variates conforms to the χ^2 distribution (2) of § 74. (See Fisher, 1935, 1, pp. 353–4.)

4. *Geometrical proof of the χ^2 distribution.* The following geometrical proof, that the sum of the squares of n independent standard normal variates x_i has a distribution given by (2) of § 74, is due to Fisher (1935, 1, p. 354). As in § 74 the probability that the values of the variates will fall simultaneously in the respective intervals dx_i is

$$dp = (2\pi)^{-\frac{1}{2}n} \exp(-\tfrac{1}{2}\chi^2)\, dx_1 dx_2 \ldots dx_n, \tag{i}$$

where

$$\chi^2 = \sum_i x_i^2.$$

Let us regard the x_i as coordinates of the current point P in Euclidean space of n dimensions. Then dp is the probability that P will fall in the element of volume $dx_1 dx_2 \ldots dx_n$. The coefficient of this element of volume in (i) is therefore the probability density for this space, and it is proportional to $\exp(-\tfrac{1}{2}\chi^2)$. Now we can express, in terms of χ and $d\chi$, an element of volume in which the value of χ may be regarded as constant. For, since χ^2 is the square of the distance of P from the origin, χ is constant over the surface of a hypersphere with radius χ and centre at the origin. The volume enclosed by this hypersphere is proportional to χ^n; and the element of volume between this and the adjacent hypersphere of radius $\chi + d\chi$ is proportional to $d(\chi^n)$, that is to $\chi^{n-1} d\chi$. Using the above value of the probability density, we see that the probability that P will fall in the region bounded by the two hyperspheres is proportional to $\chi^{n-1} \exp(-\tfrac{1}{2}\chi^2)\, d\chi$; and this is the probability that the value of χ from the random sample will fall in the interval $d\chi$. The above probability is clearly proportional to

$$(\tfrac{1}{2}\chi^2)^{\frac{1}{2}(n-2)} \exp(-\tfrac{1}{2}\chi^2)\, d(\tfrac{1}{2}\chi^2),$$

and, since χ^2 must lie between 0 and $+\infty$, the constant factor is $1/\Gamma(\tfrac{1}{2}n)$, so that the integral of the probability throughout this range may be unity. Since χ^2 lies in the interval $d\chi^2$ when χ lies in the interval $d\chi$, it follows that the distribution of χ^2 is given by § 74 (2).

For still another proof of the χ^2 distribution see Plummer, 1940, 1, p. 246.

5. *Homogeneity of several estimates of the population variance.* Suppose that k independent samples have furnished estimates s_i^2 of the population variance, based on ν_i $(i = 1, ..., k)$, D.F. Are these estimates such that the samples may be regarded as drawn from the same population? In other words, are the estimates *homogeneous*?

On the hypothesis that the samples are from the same population, an unbiased estimate s^2 of the variance of the population is

$$s^2 = \frac{1}{\nu} \sum_i \nu_i s_i^2 \quad (\nu = \sum_i \nu_i),$$

since the expected value of this quantity is the variance of the population, in virtue of §54(8). Bartlett* has shown that the statistic

$$\frac{1}{C} (\nu \log_e s^2 - \sum_i \nu_i \log_e s_i^2),$$

where

$$C = 1 + \frac{1}{3(k-1)} \left(\sum \frac{1}{\nu_i} - \frac{1}{\nu} \right),$$

is distributed approximately as χ^2 with $k-1$ D.F. The value of χ^2 calculated from the data will tell whether the hypothesis of homogeneity is reasonable.

6. Show that the estimates 3·8, 4·4, 8·1, 6·1, and 9·4 of the population variance, based on 5, 8, 6, 7 and 4 D.F. respectively, may be regarded as homogeneous according to the test of Ex. 5 $(k = 5, \chi^2 = 1·45)$.

7. That the sum $\sum_i n_i(\bar{y}_i - \bar{y})^2/\sigma_2^2$ of §83 is distributed like χ^2 with $h-1$ D.F. may also be shown as follows. Since $\bar{y}$ is the weighted mean of the $\bar{y}_i$ we have

$$\sum n_i(\bar{y}_i - \mu')^2 = \sum n_i(\bar{y}_i - \bar{y})^2 + n(\bar{y} - \mu')^2. \tag{ii}$$

Now $\bar{y}_i - \mu'$ is distributed normally about zero as mean with variance σ_2^2/n_i. Consequently $n_i(\bar{y}_i - \mu')^2/\sigma_2^2$ is a χ^2 with 1 D.F.; and,

* *Proc. Roy. Soc.* A, vol. 160, 1937, pp. 268–82. See also Neyman and Pearson, 1931, 3.

on summing for all the arrays, we see that the first member of (ii) divided by σ_2^2 is a χ^2 with h D.F. Similarly, $\bar{y} - \mu'$ is distributed normally about zero as mean with variance σ_2^2/n; and the last term in (ii), divided by σ_2^2, is therefore a χ^2 with 1 D.F. Theorem II of § 81 then shows that the first sum on the right of (ii), divided by σ_2^2, is a χ^2 with $h-1$ D.F., as stated.

8. Taking the correlation ratio as positive, deduce from § 83 (31) that the mean value of the correlation ratio of y on x, for samples from an uncorrelated bivariate normal population, is

$$\Gamma(\tfrac{1}{2}h)\, \Gamma(\tfrac{1}{2}(n-1))/\Gamma(\tfrac{1}{2}(h-1))\, \Gamma(\tfrac{1}{2}n).$$

9. Integrating by parts show that, if r is an *even* positive integer,

$$\int_\beta^\infty \frac{1}{r!} e^{-t} t^r \, dt = \frac{1}{r!} \beta^r e^{-\beta} + \int_\beta^\infty \frac{1}{(r-1)!} e^{-t} t^{r-1} \, dt,$$

and, by continuing the process, show that the value of the integral is

$$e^{-\beta} \left(1 + \beta + \frac{\beta^2}{2!} + \frac{\beta^3}{3!} + \dots + \frac{\beta^r}{r!} \right).$$

Hence show that the probability, in random sampling, that the value of χ^2 with ν (even) D.F. will exceed χ_0^2, is obtained from this expression by putting $\beta = \tfrac{1}{2}\chi_0^2$ and $r = \tfrac{1}{2}(\nu - 2)$. (Cf. Fisher, 1935, 1, pp. 356–7.)

10. The variables x, y are normally correlated with coefficient ρ. Show that u and v, defined by

$$u = x/\sigma_1 + y/\sigma_2, \quad v = x/\sigma_1 - y/\sigma_2,$$

are independent normal variates with variances $2(1+\rho)$ and $2(1-\rho)$ respectively; and that, if R is the correlation between u and v in random samples of n pairs from the bivariate population, R^2 is a $\beta_1(\tfrac{1}{2}, \tfrac{1}{2}n - 1)$ variate.

CHAPTER X

FURTHER TESTS OF SIGNIFICANCE.
SMALL SAMPLES

84. Small samples

The use of the standard errors of statistics in the tests of Chapters
VI and VII depends upon the fact that, in the case of large samples,
the sampling distributions of many statistics are approximately
normal, or at any rate unimodal with the property that a value of
the statistic, deviating from its mean by more than two or three
times its S.E., is very unlikely. For small samples, however, the
distributions of statistics are often far from normal. Moreover,
the estimate of a parameter of the population made from a small
sample is not at all reliable. For these reasons the use of standard
errors in connection with such samples is very limited. The chief
concern of the theory of small samples is with the distributions of
various statistics, and the applications of tests of significance based
upon these distributions. In each application we test a hypothesis
concerning the source of the sample. This hypothesis may be, for
instance, that a certain parameter of the population has a specified
value, or that two given samples were drawn from the same population. In each instance the test employed enables us to form a con-
clusion based on considerations of probability. The nature of the
tests is illustrated by the χ^2 test of § 78, which is applicable to
samples of any size. But, as already indicated, the test of goodness
of fit considered in § 79 can be applied only to large samples.

We remind the reader of two important results obtained earlier.
First, in simple sampling from a population with mean μ and
variance σ^2, the distribution of the mean $\bar{x}$ of the sample of n
members has μ for its mean and σ^2/n for its variance (see § 50).
This result holds whether the sample is large or small, and whether
the population is normal or not. In the case of a normal population
the distribution of $\bar{x}$ is also normal (see §§ 22 or 82). Secondly, the
statistic s^2 defined by

$$(n-1)s^2 = \sum_{i=1}^{n} (x_i - \bar{x})^2 \tag{1}$$

is an unbiased estimate of the population variance σ^2, in the sense that $E(s^2) = \sigma^2$ (cf. § 54 (8)). This estimate of σ^2 is said to be based on $n-1$ D.F., since the variates $x_i - \bar{x}$ are not independent, being connected by the linear relation $\sum(x_i - \bar{x}) = 0$. The number of independent variates is thus only $n-1$; and this agrees with the result of § 77, that $(n-1)s^2/\sigma^2$ is distributed like χ^2 with $n-1$ D.F.

The following discussion applies to *samples of any size*. It is assumed, however, that the populations from which the samples are drawn are normal. The results obtained are therefore strictly true only in this case. But they are approximately true, and may be usefully applied, in most cases in which the departure of the population from normality is not very marked.

'STUDENT'S' DISTRIBUTION

85. The statistic *t* and its distribution

We have seen that, in simple sampling from a normal population of mean μ and variance σ^2, the deviation $\bar{x} - \mu$ is a normal variate, with mean zero and s.d. $\sigma/\sqrt{n}$. The quotient of $\bar{x} - \mu$ by $\sigma/\sqrt{n}$ is therefore distributed normally with unit s.d. If, however, in place of the constant σ we use the variable estimate s obtained from the sample, we have the statistic

$$t = \frac{(\bar{x} - \mu)\sqrt{n}}{s}, \tag{2}$$

which is not normally distributed. The distribution of t was first found by W. S. Gosset, who wrote under the *nom de plume* of 'Student'. It follows very simply from the theorems of §§ 67, 73 and 77. For, from its definition,

$$t = \frac{(\bar{x} - \mu)\sqrt{n}}{\sqrt{(nS^2/\nu)}},$$

where $nS^2 = \sum(x_i - \bar{x})^2$ and ν is the number of D.F., $n-1$, on which the estimate s^2 is based. Hence

$$\frac{t^2}{\nu} = \frac{\frac{1}{2}(\bar{x} - \mu)^2 \div (\sigma^2/n)}{\frac{1}{2}nS^2/\sigma^2}.$$

But, in virtue of §67 (Theorem I) and §77, the numerator in the second member is a Gamma variate with parameter $\frac{1}{2}$, and the denominator a Gamma variate with parameter $\frac{1}{2}\nu$; and these are distributed independently of each other (cf. §82 or Ex. VIII, 11). Consequently t^2/ν is a $\beta_2(\frac{1}{2}, \frac{1}{2}\nu)$ variate; and its distribution is obtained by substituting t^2/ν for x in §72 (27), with $l = \frac{1}{2}$ and $m = \frac{1}{2}\nu$. Thus the probability that, in random sampling, the value of t^2 will fall in the interval dt^2 is

$$dP = \frac{(t^2)^{-\frac{1}{2}} d(t^2)}{\sqrt{\nu} \cdot B(\frac{1}{2}, \frac{1}{2}\nu)(1 + t^2/\nu)^{\frac{1}{2}(\nu+1)}}. \tag{3}$$

By integrating with respect to t^2 from a fixed value t_0^2 to infinity,[*] we obtain the probability that the value of t^2 will exceed t_0^2. The accompanying table contains extracts from a more complete one given by Fisher. In the body of the table are given the values of t_0 corresponding to certain fixed values of ν and P. Thus, for a specified number ν of D.F., the value of P at the top of the column is the probability that, in random sampling, the numerical value of t in the body of the table will be exceeded.

The range of t^2 is from 0 to ∞, and its distribution is given by (3). The statistic t, however, ranges from $-\infty$ to $+\infty$, and its distribution is therefore

$$dP = \frac{dt}{\sqrt{\nu} \cdot B(\frac{1}{2}, \frac{1}{2}\nu)(1 + t^2/\nu)^{\frac{1}{2}(\nu+1)}}, \tag{4}$$

the factor 2 disappearing, since the integral of dP over the whole range of variation of t must be unity. The distribution (4) is spoken of as the t distribution corresponding to ν D.F. And from the above argument it is clear that any statistic t, whose range is from $-\infty$ to $+\infty$, and which is such that t^2/ν is a $\beta_2(\frac{1}{2}, \frac{1}{2}\nu)$ variate, conforms to the t distribution for ν D.F. This important result may also be expressed in the form of

THEOREM I. *A statistic t conforms to the t distribution for ν D.F. if its range is from $-\infty$ to $+\infty$, and t^2/ν is expressible as the quotient of two independent variates, which are distributed like χ^2 with 1 and ν D.F. respectively.*

---

[*] For a method of evaluating this integral see Fisher, 1935, 1, pp. 358–60.

TABLE 4. *Values of mod. t with a probability P
of being exceeded in random sampling*

ν = number of degrees of freedom

ν \ P	0·50	0·10	0·05	0·02	0·01
1	1·000	6·34	12·71	31·82	63·66
2	0·816	2·92	4·30	6·96	9·92
3	0·765	2·35	3·18	4·54	5·84
4	0·741	2·13	2·78	3·75	4·60
5	0·727	2·02	2·57	3·36	4·03
6	0·718	1·94	2·45	3·14	3·71
7	0·711	1·90	2·36	3·00	3·50
8	0·706	1·86	2·31	2·90	3·36
9	0·703	1·83	2·26	2·82	3·25
10	0·700	1·81	2·23	2·76	3·17
11	0·697	1·80	2·20	2·72	3·11
12	0·695	1·78	2·18	2·68	3·06
13	0·694	1·77	2·16	2·65	3·01
14	0·692	1·76	2·14	2·62	2·98
15	0·691	1·75	2·13	2·60	2·95
16	0·690	1·75	2·12	2·58	2·92
17	0·689	1·74	2·11	2·57	2·90
18	0·688	1·73	2·10	2·55	2·88
19	0·688	1·73	2·09	2·54	2·86
20	0·687	1·72	2·09	2·53	2·84
21	0·686	1·72	2·08	2·52	2·83
22	0·686	1·72	2·07	2·51	2·82
23	0·685	1·71	2·07	2·50	2·81
24	0·685	1·71	2·06	2·49	2·80
25	0·684	1·71	2·06	2·48	2·79
26	0·684	1·71	2·06	2·48	2·78
27	0·684	1·70	2·05	2·47	2·77
28	0·683	1·70	2·05	2·47	2·76
29	0·683	1·70	2·04	2·46	2·76
30	0·683	1·70	2·04	2·46	2·75
35	0·682	1·69	2·03	2·44	2·72
40	0·681	1·68	2·02	2·42	2·71
45	0·680	1·68	2·02	2·41	2·69
50	0·679	1·68	2·01	2·40	2·68
60	0·678	1·67	2·00	2·39	2·66
∞	0·674	1·64	1·96	2·33	2·58

Reproduced by permission of the author, Professor R. A. Fisher, from his book
on *Statistical Methods for Research Workers.*

The reader should sketch the probability curves for t^2 and t.
The former is asymptotic to both axes, with ordinate which decreases
continuously as t^2 increases. The probability curve of t is symmetrical
about the line $t = 0$. It is asymptotic to the t-axis at each end, and

the maximum ordinate is that for $t = 0$. Thus small values of t^2 are more likely than larger values. In this respect t^2 differs from χ^2 when the latter has more than 2 D.F.

86. Test for an assumed population mean

The statistic t and the above table provide a means of testing an assumed value μ for the mean of the normal population from which the random sample was drawn. On the hypothesis that μ is the true value, the equation (2) and the values of $\bar{x}$ and s derived from the sample enable us to calculate t. The table then gives the probability P that this value will be exceeded numerically in random sampling from a normal population with mean μ. If P is less than 0·05 we regard our value of t as significant. If P is less than 0·01 we regard it as highly significant. A significant value of t throws doubt on the truth of the hypothesis that μ is the mean of the population.

Example. A random sample of nine from the men of a large city gave a mean height of 68 in.; and the unbiased estimate s^2 of the population variance found from the sample was 4·5 in.[2] Are these data consistent with the assumption of a mean height of 68·5 in. for the men of the city?

Large populations of heights of men are known to be approximately normal. For the given sample

$$\bar{x} = 68{\cdot}0, \quad \nu = 9 - 1 = 8, \quad s = \sqrt{4{\cdot}5} = 2{\cdot}12,$$

and therefore, on the hypothesis that $\mu = 68{\cdot}5$, we have

$$|t| = |(\bar{x} - \mu)|\sqrt{n}/s = (0{\cdot}5) \times 3/2{\cdot}12 = 0{\cdot}707.$$

From the table we find that, for $\nu = 8$, the probability that this value of t will be exceeded numerically in random sampling is about 0·50. The value is therefore not at all significant, and the test provides no evidence against the assumption of a population mean of 68·5 in.

If we make the assumption that $\mu = 69{\cdot}5$ or 66·5 in. we obtain a value of $|t|$ three times as great as before, viz. 2·12. The probability that, for $\nu = 8$, this value will be exceeded in random sampling is greater than 0·05, and the new value of t is still not significant. There are thus fairly wide limits for the assumed population mean which, with the data of the sample, will provide a value of t which is not significant. These limits we proceed to consider.

87. Fiducial limits* for the population mean

Suppose that a certain sample from a normal population has a mean $\bar{x}$, and provides an unbiased estimate s^2 of the population

* Cf. Fisher, 1930, 3 and 1933, 1.

variance based on ν D.F. We wish to find limits to the assumed population mean μ so that, with the data of the sample, it will lead to a value of t that is not significant. If our choice is the 5 % level of significance, we define the 95 % confidence range for μ as that range of values of μ which, with the data of the sample, will furnish a value of $|t|$ less than the value t_1 which corresponds to $P = 0.05$. This requires

$$|\bar{x} - \mu| \sqrt{n}/s < t_1,$$

so that $\qquad \bar{x} - st_1/\sqrt{n} < \mu < \bar{x} + st_1/\sqrt{n}. \qquad (5)$

Consequently μ must lie within the range extending from $\bar{x} - st_1/\sqrt{n}$ to $\bar{x} + st_1/\sqrt{n}$, which is called the 95 % confidence range for μ corresponding to the given sample; and the bounding values of this range are the corresponding confidence limits, or *fiducial limits*, for the mean of the population. Similarly we have confidence ranges and limits corresponding to other levels of significance. In each case the appropriate value of t_1 is found from the table of t.

Example 1. Find the 95 % fiducial limits for μ corresponding to the sample in the example of § 86.

Here $\nu = 8$, $t_1 = 2.31$, $s = 2.12$, $n = 9$. Hence

$$st_1/\sqrt{n} = (2.12)(2.31)/3 = 1.63.$$

The required limits are therefore 68 ± 1.63, that is 66.37 and 69.63 in.

Example 2. Find the 98 % fiducial limits for μ corresponding to the same sample.

From the table we find that $t_1 = 2.90$ is the value of t which is exceeded numerically with a probability of 2 %. Then

$$st_1/\sqrt{n} = (2.12)(2.90)/3 = 2.05,$$

and the required limits are 68 ± 2.05, that is 65.95 and 70.05 in.

88. Comparison of the means of two samples

Given two independent samples of n_1 and n_2 members, with means $\bar{x}_1$ and $\bar{x}_2$ respectively, we may use the t distribution to decide whether the means differ significantly, or whether the two samples may be regarded as drawn from the same normal population.* We test the hypothesis that they are from the same normal population.

* Cf. Fisher, 1925, 1, pp. 90–3.

Let x_i $(i = 1, ..., n_1)$ be the values of the variable in the first sample, and x'_j $(j = 1, ..., n_2)$ those in the second. Then the sums of squares for the two samples are

$$n_1 S_1^2 = \sum_i (x_i - \bar{x}_1)^2, \quad n_2 S_2^2 = \sum_j (x'_j - \bar{x}_2)^2$$

respectively. Also, if σ^2 is the variance of the population, $n_1 S_1^2/\sigma^2$ and $n_2 S_2^2/\sigma^2$ are distributed like χ^2 with $n_1 - 1$ and $n_2 - 1$ D.F. respectively. Consequently $(n_1 S_1^2 + n_2 S_2^2)/\sigma^2$ is a χ^2 with ν D.F., where

$$\nu = n_1 + n_2 - 2.$$

An unbiased estimate of the population variance, obtained from the samples, is

$$s^2 = (n_1 S_1^2 + n_2 S_2^2)/\nu,$$

since

$$E(\nu s^2) = E(n_1 S_1^2 + n_2 S_2^2) = (n_1 - 1)\sigma^2 + (n_2 - 1)\sigma^2 = \nu\sigma^2,$$

so that

$$E(s^2) = \sigma^2.$$

Now, in virtue of the hypothesis, $\bar{x}_1$ and $\bar{x}_2$ are normally distributed about the population mean with variances σ^2/n_1 and σ^2/n_2 respectively. Therefore, since the samples are independent, the difference $\bar{x}_1 - \bar{x}_2$ is normally distributed about zero with variance $\sigma^2(1/n_1 + 1/n_2)$. If then we define a statistic t by the equation

$$t = \frac{\bar{x}_1 - \bar{x}_2}{s\sqrt{(1/n_1 + 1/n_2)}}, \tag{6}$$

we have

$$\frac{t^2}{\nu} = \frac{(\bar{x}_1 - \bar{x}_2)^2/\sigma^2(1/n_1 + 1/n_2)}{(n_1 S_1^2 + n_2 S_2^2)/\sigma^2}.$$

The numerator and denominator of the second member are distributed independently like χ^2 with 1 and ν D.F. respectively. Hence, by the Theorem of § 85, the statistic t conforms to the t distribution for ν D.F. It is the quotient of the normal deviate $\bar{x}_1 - \bar{x}_2$ by the estimate of its S.E. derived from the samples. If the value of t obtained from the samples is significant, our hypothesis is discredited.

We might vary the above argument to test the hypothesis that the samples were drawn from different normal populations, with means μ and μ' respectively, but *the same variance*. In this case $\bar{x}_1 - \mu$ and $\bar{x}_2 - \mu'$ are normally distributed about zero with variances σ^2/n_1 and σ^2/n_2 respectively. Hence their difference

$$(\bar{x}_1 - \bar{x}_2) - (\mu - \mu')$$

is normally distributed about zero with variance $\sigma^2(1/n_1 + 1/n_2)$. This difference takes the place of $\bar{x}_1 - \bar{x}_2$ in the above argument, so that

$$t = \frac{(\bar{x}_1 - \bar{x}_2) - (\mu - \mu')}{s\sqrt{(1/n_1 + 1/n_2)}} \tag{7}$$

conforms to the t distribution for ν D.F.

Example 1. Show that the 95 % fiducial limits for the difference of the means of the populations are $\bar{x}_1 - \bar{x}_2 \pm t_1 s\sqrt{(1/n_1 + 1/n_2)}$, where t_1 is the value of t corresponding to $P = 0.05$ and ν D.F. If zero lies outside the above range of $\mu - \mu'$, we conclude that the difference between the means of the samples is significant of a difference between the means of the populations.

Example 2. The heights of the ten men of a random sample from an unknown population gave a mean of 69 in., and a sum of squares of deviations from the mean equal to 42 in.[2] Apply the t test to the hypothesis that this sample is from the same population as that of the example in § 86.

From the data we have

$$\bar{x}_1 = 68, \quad \bar{x}_2 = 69, \quad n_1 S_1^2 = 36, \quad n_2 S_2^2 = 42, \quad n_1 = 9, \quad n_2 = 10.$$

The estimate s^2 of the population variance is

$$s^2 = (36 + 42)/17 = 4.59,$$

so that $s = 2.14$. The value of t from the sample is then

$$t = \frac{1}{2.14}\sqrt{\frac{90}{19}} = 1.01.$$

For $\nu = 17$ this value is not at all significant. The test therefore provides no evidence against the hypothesis.

89. Significance of an observed correlation

The distribution and table of t may also be used to test the significance of a value r of the correlation coefficient, given by a random sample of n pairs of values from a bivariate normal population. The s.e. of r may not be used in the case of small samples, since the distribution of r is then far from normal. We have seen that, if the variables in the normal population are uncorrelated, the value of r^2 in random samples of n pairs is a $\beta_1(\tfrac{1}{2}, \tfrac{1}{2}(n-2))$

variate. Therefore, by Theorem VII of § 72, $r^2/(1-r^2)$ is a $\beta_2(\frac{1}{2}, \frac{1}{2}(n-2))$ variate. Thus, if t is defined by

$$t = \frac{r}{\sqrt{(1-r^2)}}\sqrt{(n-2)}, \tag{8}$$

then $t^2/(n-2)$ is a $\beta_2(\frac{1}{2}, \frac{1}{2}(n-2))$ variate, so that t conforms to the t distribution for $n-2$ D.F.

To test the significance of the sample value of r, we make the assumption that the variables in the population are uncorrelated. A simple calculation then gives the value of t corresponding to the sample; and the table of t then tells whether the value obtained is a rare one. If it is, our assumption is discredited, and we conclude that the variables in the population are probably correlated.

Example 1. Is a correlation coefficient of 0·5 significant, if obtained from a random sample of 11 pairs of values from a normal population?

Here $r = \frac{1}{2}$, $\nu = 9$, $t = \frac{1}{2} \times 3 \div \sqrt{(3/4)} = \sqrt{3} = 1·73$. From the table we find that the probability of obtaining a value of t larger than this is greater than 0·10. Hence there is no reason to suspect the hypothesis of uncorrelated variables in the population. The value 0·5 is not significant.

Example 2. Find the least value of r, in a sample of 27 pairs from a normal population, that is significant at the 5 % level.

Here $\nu = 25$ and, at the 5 % level of significance, $t = 2·06$. Hence for r to be significant we must have

$$\frac{5r}{\sqrt{(1-r^2)}} > 2·06,$$

which requires $r^2 > 0·145$ and therefore $|r| > 0·38$. Values of r numerically less than 0·38 are not significant at the 5 % level.

The accompanying short table gives the least values of $|r|$ that

TABLE 5. *Minimum values of r that are significant at the 5 % level*

ν	r	ν	r	ν	r	ν	r
4	0·811	11	0·553	18	0·444	45	0·288
5	0·755	12	0·532	19	0·433	50	0·273
6	0·707	13	0·514	20	0·423	60	0·250
7	0·666	14	0·497	25	0·381	70	0·232
8	0·632	15	0·482	30	0·349	80	0·217
9	0·602	16	0·468	35	0·325	90	0·205
10	0·576	17	0·456	40	0·304	100	0·195

Reproduced by permission of the author, Professor R. A. Fisher, from his book on *Statistical Methods for Research Workers*.

are significant at the 5 % level, for samples of different sizes from a normal population. These values may be calculated as in Ex. 2 above. The number ν of degrees of freedom is $n-2$, where n is the number of pairs of values in the sample.

90. Significance of an observed regression coefficient

The significance of an observed value, b, of the linear regression coefficient of y on x, in a random sample from a normal population, may also be tested by the t distribution and table. It was shown in § 83 (Theorem V) that, for random samples of n pairs from an un-correlated normal population, $b^2\sigma_1^2/\sigma_2^2$ is a $\beta_2(\frac{1}{2}, \frac{1}{2}(n-1))$ variate. Consequently the statistic t defined by

$$t = b\sigma_1 \sqrt{(n-1)}/\sigma_2, \tag{9}$$

conforms to the t distribution for $n-1$ D.F. But, since σ_1 and σ_2 are usually unknown, this relation is not of much use in providing a test of significance for b. However, a similar result is free from this objection. For, from the relation

$$b = rS_2/S_1, \tag{10}$$

in which S_1^2 and S_2^2 are the variances of x and y in the sample, it follows that

$$\frac{b^2 \sum (x_i - \bar{x})^2}{\Sigma(y_i - Y_i)^2} = \frac{nr^2 S_2^2/\sigma_2^2}{\sum (y_i - Y_i)^2/\sigma_2^2},$$

where Y_i as usual denotes the estimate of y_i found from the regression equation. Now, by § 82, the numerator and denominator of the second member are distributed like χ^2 with 1 and $n-2$ D.F. respectively. The quotient is therefore a $\beta_2(\frac{1}{2}, \frac{1}{2}(n-2))$ variate, and the statistic t defined by

$$t = b \sqrt{\{(n-2) \sum (x_i - \bar{x})^2/ \sum (y_i - Y_i)^2\}}, \tag{11}$$

conforms to the t distribution for $n-2$ D.F., and may be used to test the significance of the value of b found from the sample. More generally Fisher* has shown that, if the variables in the population are correlated, with β as coefficient of regression of y on x, the statistic which conforms to the t distribution is obtained from (11) on replacing b by $(b-\beta)$.

* Fisher, 1922, 2, p. 609.

91. Distribution of the range of a sample.

The range, w, of a sample of values is the difference between the highest and lowest values. This concept is widely used in the statistical control of quality in mass production by an industrial plant. Though the range does not conform to the t-distribution, it is convenient here to consider the sampling distribution of w for samples of n values from a continuous population, whose relative frequency density is $f(x)$ in the interval (a, b). Consider the infinitesimal intervals $(u, u+du)$ and $(v, v+dv)$ within the range of variation of x, and such that $u < v$. Then the probability that, at any drawing, the value chosen will lie in the first of these is $f(u)\,du$, and in the second $f(v)\,dv$. The probability that it will lie in the intervening interval is

$$p = \int_{u+du}^{v} f(x)\,dx. \tag{i}$$

Consequently the probability that, in the drawing of n values, one will lie in the interval du, one in the interval dv, and the remaining $n-2$ in the intervening interval is

$$dP = n(n-1)f(u)f(v)\,p^{n-2}\,du\,dv, \tag{ii}$$

since $n(n-1)$ is the number of ways in which one of the n values may fall in each of the intervals du and dv. In other words, dP is the probability that the lowest value in the sample will fall in the interval du, and highest in the interval dv.

We may express this in terms of u and w instead of u and v. Since $w = v - u$ the Jacobian of u, w with respect to u, v is unity, and the probability that the lowest value will fall in the interval du, and w in the interval dw, is therefore

$$dP' = n(n-1)f(u)f(u+w)\left(\int_{u}^{u+w} f(x)\,dx\right)^{n-2} du\,dw \tag{iii}$$

to within infinitesimals of this order. Summing for all intervals du consistent with a range w, we find the probability that w will fall in the interval dw, irrespective of the value of u, as

$$dp = n(n-1)\left[\int_{a}^{b-w} f(u)f(u+w)\left(\int_{u}^{u+w} f(x)\,dx\right)^{n-2} du\right]dw. \tag{12}$$

This is the required probability differential of w. Denoting the coefficient of dw by $\phi(w)$ we have the expected value of w as

$$E(w) = \int_0^{b-a} w\phi(w)\, dw.$$

This expected value has been calculated by numerical integration for the case of a *normal* population of S.D. σ, and for various values of n. Among the results found are

n:	2	3	4	5	6	10	50
$E(w)/\sigma$:	1·13	1·69	2·06	2·33	2·53	3·08	4·50

Example 1. If the variable in the population has a uniform distribution from 0 to b, then $f(x) = 1/b$. Show that the probability differential of the range is $n(n-1)\, b^{-n} w^{n-2} (b-w)\, dw$, and that the expected value of w is $(n-1)\, b/(n+1)$.

Example 2. Show that, in taking a random sample of n numbers between zero and unity, the probability that the range will exceed 0·5 is $1 - (n+1)/2^n$. Show also that 8 is the least value of n for which this probability exceeds 0·95.

DISTRIBUTION OF THE VARIANCE RATIO

92. Ratio of independent estimates of the population variance

As in § 88, let us consider two independent random samples whose values are x_i ($i = 1, ..., n_1$), and x'_j ($j = 1, ..., n_2$), with means $\bar{x}_1$ and $\bar{x}_2$ respectively. These provide estimates s_1^2 and s_2^2 of the variances of the populations, given by

$$s_1^2 = \frac{\sum (x_i - \bar{x}_1)^2}{n_1 - 1} = \frac{n_1 S_1^2}{\nu_1}, \quad s_2^2 = \frac{\sum (x'_j - \bar{x}_2)^2}{n_2 - 1} = \frac{n_2 S_2^2}{\nu_2}.$$

corresponding to ν_1 and ν_2 D.F. respectively, where

$$\nu_1 = n_1 - 1, \quad \nu_2 = n_2 - 1.$$

We wish to consider whether two such estimates are significantly different, or whether the samples may be regarded as drawn from the same normal population of variance σ^2.

An appropriate test is furnished by the sampling distribution of the ratio F of two such estimates of σ^2 obtained from independent samples from the same normal population. Thus if

$$F = \frac{s_1^2}{s_2^2} = \frac{n_1 S_1^2/\nu_1}{n_2 S_2^2/\nu_2}, \tag{13}$$

then

$$\frac{\nu_1 F}{\nu_2} = \frac{n_1 S_1^2/\sigma^2}{n_2 S_2^2/\sigma^2}.$$

Now the numerator and denominator of the second member are distributed independently like χ^2 with ν_1 and ν_2 D.F. respectively. Hence the quotient is a $\beta_2(\tfrac{1}{2}\nu_1, \tfrac{1}{2}\nu_2)$ variate, so that $\nu_1 F/\nu_2$ conforms to the distribution of § 72, with $l = \tfrac{1}{2}\nu_1$ and $m = \tfrac{1}{2}\nu_2$. Consequently the probability that the value of F will fall in the interval dF is

$$dP = \frac{\nu_1^{\frac{1}{2}\nu_1}\nu_2^{\frac{1}{2}\nu_2} F^{\frac{1}{2}(\nu_1-2)} dF}{B(\tfrac{1}{2}\nu_1, \tfrac{1}{2}\nu_2)(\nu_1 F + \nu_2)^{\frac{1}{2}(\nu_1+\nu_2)}}. \tag{14}$$

This is the required distribution of the variance ratio for ν_1 and ν_2 D.F. It will be observed that the distribution is independent of the variance σ^2 of the population.

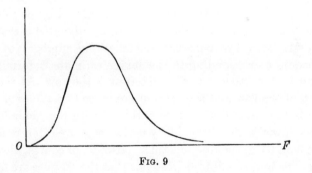

FIG. 9

Since $\nu_1 F/\nu_2$ is a $\beta_2(\tfrac{1}{2}\nu_1, \tfrac{1}{2}\nu_2)$ variate its modal value, by § 72 (28), is $(\tfrac{1}{2}\nu_1 - 1)/(\tfrac{1}{2}\nu_2 + 1)$. Consequently the modal value of F is given by

$$\breve{F} = \frac{\nu_2(\nu_1 - 2)}{\nu_1(\nu_2 + 2)} = \frac{\nu_1\nu_2 - 2\nu_2}{\nu_1\nu_2 + 2\nu_1},$$

and this is always *less than unity*. Similarly, the mean value of $\nu_1 F / \nu_2$, by § 72 (29), is $\frac{1}{2}\nu_1 / (\frac{1}{2}\nu_2 - 1)$; and therefore the expected value of F is

$$E(F) = \frac{\nu_2}{\nu_2 - 2},$$

which is *independent of* ν_1, and is always *greater than unity*. The probability curve for F depends, of course, on both ν_1 and ν_2; but its main features, for $\nu_1 > 4$, are those of the curve in Fig. 9.

93. Fisher's z distribution. Table of F

A distribution equivalent to (14) was first obtained by R. A. Fisher. Writing $z = \frac{1}{2}\log_e F$ and therefore $F = e^{2z}$ in the above result, we deduce immediately that

$$dP = \frac{2\nu_1^{\frac{1}{2}\nu_1}\nu_2^{\frac{1}{2}\nu_2} e^{\nu_1 z}\, dz}{B(\frac{1}{2}\nu_1, \frac{1}{2}\nu_2)\,(\nu_1 e^{2z} + \nu_2)^{\frac{1}{2}(\nu_1 + \nu_2)}}. \tag{15}$$

This is Fisher's z distribution. The probability that a specified value of z will be exceeded in random sampling depends upon ν_1 and ν_2. Fisher published tables* giving the values of z that will be exceeded with probabilities 0·05 and 0·01 respectively, corresponding to specified values of ν_1 and ν_2. From these G. W. Snedecor prepared a table† for the variance ratio, which he denoted by F in honour of Fisher. Extracts from this table are here printed by permission of Snedecor and the Iowa Press. The ratio, F, tabulated is that of the larger estimates of variance to the smaller. The number ν_1 of degrees of freedom corresponding to the larger estimate determines the column in the table, while ν_2 determines the row. At the intersection of the row and the column are given two values of F. The upper is the value that will be exceeded with a probability 0·05, and the lower with a probability 0·01. These are often referred to as the 5 and 1 % 'points' of F. The latter is, of course, always the larger. The hypothesis to be tested is that the samples are from the same normal population, or from normal populations of equal variance. A value of F less than the 5 % point is not significant. A value between the 5 and 1 % points is significant at the former

---

* *Statistical Methods for Research Workers*, 1925.

† Snedecor, 1934, 2 or 1938, 3, pp. 184–7.

TABLE 6. *The variance ratio.* 5 *and* 1 % '*points*' *of F*

ν_1 is the number of degrees of freedom for the greater
estimate of variance, and ν_2 for the smaller

ν_1 ν_2	1	2	3	4	5	6	8	12	24	∞
2	18·51	19·00	19·16	19·25	19·30	19·33	19·37	19·41	19·45	19·50
	98·49	99·00	99·17	99·25	99·30	99·33	99·36	99·42	99·46	99·50
3	10·13	9·55	9·28	9·12	9·01	8·94	8·84	8·74	8·64	8·53
	34·12	30·82	29·46	28·71	28·24	27·91	27·49	27·05	26·60	26·12
4	7·71	6·94	6·59	6·39	6·26	6·16	6·04	5·91	5·77	5·63
	21·20	18·00	16·69	15·98	15·52	15·21	14·80	14·37	13·93	13·46
5	6·61	5·79	5·41	5·19	5·05	4·95	4·82	4·68	4·53	4·36
	16·26	13·27	12·06	11·39	10·97	10·67	10·27	9·89	9·47	9·02
6	5·99	5·14	4·76	4·53	4·39	4·28	4·15	4·00	3·84	3·67
	13·74	10·92	9·78	9·15	8·75	8·47	8·10	7·72	7·31	6·88
7	5·59	4·74	4·35	4·12	3·97	3·87	3·73	3·57	3·41	3·23
	12·25	9·55	8·45	7·85	7·46	7·19	6·84	6·47	6·07	5·65
8	5·32	4·46	4·07	3·84	3·69	3·58	3·44	3·28	3·12	2·93
	11·26	8·65	7·59	7·01	6·63	6·37	6·03	5·67	5·28	4·86
9	5·12	4·26	3·86	3·63	3·48	3·37	3·23	3·07	2·90	2·71
	10·56	8·02	6·99	6·42	6·06	5·80	5·47	5·11	4·73	4·31
10	4·96	4·10	3·71	3·48	3·33	3·22	3·07	2·91	2·74	2·54
	10·04	7·56	6·55	5·99	5·64	5·39	5·06	4·71	4·33	3·91
12	4·75	3·88	3·49	3·26	3·11	3·00	2·85	2·69	2·50	2·30
	9·33	6·93	5·95	5·41	5·06	4·82	4·50	4·16	3·78	3·36
14	4·60	3·74	3·34	3·11	2·96	2·85	2·70	2·53	2·35	2·13
	8·86	6·51	5·56	5·03	4·69	4·46	4·14	3·80	3·43	3·00
16	4·49	3·63	3·24	3·01	2·85	2·74	2·59	2·42	2·24	2·01
	8 53	6·23	5·29	4·77	4·44	4·20	3·89	3·55	3·18	2·75
18	4·41	3·55	3·16	2·93	2·77	2·66	2·51	2·34	2·15	1·92
	8·28	6·01	5·09	4·58	4·25	4·01	3·71	3·37	3·01	2·57
20	4·35	3·49	3·10	2·87	2·71	2·60	2·45	2·28	2·08	1·84
	8·10	5·85	4·94	4·43	4·10	3·87	3·56	3·23	2·86	2·42
25	4·24	3·38	2·99	2·76	2·60	2·49	2·34	2·16	1·96	1·71
	7·77	5·57	4·68	4·18	3·86	3·63	3·32	2·99	2·62	2·17
30	4·17	3·32	2·92	2·69	2·53	2·42	2·27	2·09	1·89	1·62
	7·56	5·39	4·51	4·02	3·70	3·47	3·17	2·84	2·47	2·01
40	4·08	3·23	2·84	2·61	2·45	2·34	2·18	2·00	1·79	1·51
	7·31	5·18	4·31	3·83	3·51	3·29	2·99	2·66	2·29	1·81
60	4·00	3·15	2·76	2·52	2·37	2·25	2·10	1·92	1·70	1·39
	7·08	4·98	4·13	3·65	3·34	3·12	2·82	2·50	2·12	1·60
80	3·96	3·11	2·72	2·49	2·33	2·21	2·06	1·88	1·65	1·32
	6·96	4·88	4·04	3·56	3·25	3·04	2·74	2·41	2·03	1·49

Extracted from G. W. Snedecor's *Statistical Methods*, pp. 184–7, by courtesy
of the author and the Iowa Collegiate Press.

level, but not at the latter. A value greater than the 1 % point is regarded as highly significant. A significant value of F throws doubt on the truth of the hypothesis. This test will be used extensively in the next chapter. It is illustrated by the following numerical example.

Example. Apply the test to the samples of heights given in the examples of §§ 86 and 88 (Ex. 2).

The two estimates of the population variance furnished by the samples are 36/8 and 42/9, corresponding to 8 and 9 D.F. respectively. The second is the larger, so that

$$F = \frac{42/9}{36/8} = 1.04, \quad \nu_1 = 9, \quad \nu_2 = 8.$$

This value of F is much below the 5 % point as indicated by the table. It is therefore not at all significant, so that the samples may very well be regarded as drawn from the same population.

FISHER'S TRANSFORMATION OF THE CORRELATION COEFFICIENT

94. Distribution of r. Fisher's transformation

The distribution of the correlation coefficient r, in random samples of n pairs of values from a bivariate normal population in which the correlation is ρ, was given by Fisher* in 1915. He showed that the probability that the correlation coefficient will have a value in the interval dr is

$$dp = \frac{(1-\rho^2)^{\frac{1}{2}(n-1)}}{\pi(n-3)!} (1-r^2)^{\frac{1}{2}(n-4)} \frac{d^{n-2}}{d(r\rho)^{n-2}} \left(\frac{\text{arc cos}\,(-r\rho)}{\sqrt{(1-r^2\rho^2)}} \right) dr.$$

This distribution is far from normal, with a probability curve which is very skew in the neighbourhood of $\rho = \pm 1$, even for large samples. The use of the S.E. of r is therefore not to be recommended. In a subsequent paper† Fisher showed that the transformation

$$z = \tfrac{1}{2}\log_e \frac{1+r}{1-r}, \quad \zeta = \tfrac{1}{2}\log_e \frac{1+\rho}{1-\rho}, \tag{16}$$

defines a variate z, whose distribution is approximately normal with mean ζ and variance $1/(n-3)$, tending rapidly to normality as the size of the sample increases. Thus the S.E. of z is independent of the

* Fisher, 1915, 2. † Fisher, 1921, 1.

value of *r*. For the proofs of these properties we must refer the reader to Fisher's papers; but the distribution of *r* in the case of an uncorrelated normal population was considered in § 82, and the corresponding distribution of *z* will be discussed in Ex. 1 below.

By means of the statistic *z* we may test whether an observed correlation coefficient differs significantly from some theoretical value, or from some value given in advance; or whether the values of *r* obtained from two samples differ significantly. From the values of *r* and ρ we determine those of *z* and ζ by (16); and it is then easy to decide whether the deviation $z - \zeta$ is significant for a normal distribution of variance $1/(n-3)$. To obviate the necessity of calculating *z* in every case, Fisher published a table setting out the values of *r*, which correspond to specified values of *z* ranging from 0 to 3 at intervals of 0·01. Extracts from this table are printed herewith.

TABLE 7. *Fisher's transformation of r*

Values of *r* for specified values of *z* at intervals of 0·02

z	0·00	0·02	0·04	0·06	0·08
0·0	0·000	0·020	0·040	0·060	0·080
0·1	0·100	0·119	0·139	0·159	0·178
0·2	0·197	0·217	0·236	0·254	0·273
0·3	0·291	0·310	0·328	0·345	0·363
0·4	0·380	0·397	0·414	0·430	0·446
0·5	0·462	0·478	0·493	0·508	0·523
0·6	0·537	0·551	0·565	0·578	0·592
0·7	0·604	0·617	0·629	0·641	0·653
0·8	0·664	0·675	0·686	0·696	0·706
0·9	0·716	0·726	0·735	0·744	0·753
1·0	0·762	0·770	0·778	0·786	0·793
1·1	0·801	0·808	0·814	0·821	0·828
1·2	0·834	0·840	0·846	0·851	0·857
1·3	0·862	0·867	0·872	0·876	0·881
1·4	0·885	0·890	0·894	0·898	0·902
1·5	0·905	0·909	0·912	0·915	0·919
1·6	0·922	0·925	0·928	0·930	0·933
1·7	0·936	0·938	0·940	0·943	0·945
1·8	0·947	0·949	0·951	0·953	0·955
1·9	0·956	0·958	0·960	0·961	0·963

Reproduced by permission of the author, Professor R. A. Fisher, from his book on *Statistical Methods for Research Workers*.

For the case $\zeta = 0$, that is to say for testing whether an observed value of *r* indicates any correlation in the population, the method of § 89 is preferable.

Example 1. For samples from an uncorrelated normal population, we know that the distribution of r is

$$dp = \frac{(1-r^2)^{\frac{1}{2}(n-4)}\, dr}{B(\frac{1}{2}, \frac{1}{2}(n-2))}.$$

Fisher's transformation (16) may be expressed

$$r = \tanh z,$$

so that
$$dr = \operatorname{sech}^2 z\, dz,$$

and the distribution of z is therefore

$$dp = \operatorname{sech}^{n-2} z\, dz / B(\tfrac{1}{2}, \tfrac{1}{2}(n-2)).$$

Now $\operatorname{sech} z$ is approximately equal to $\exp(-\tfrac{1}{2}z^2)$, so that

$$dp \propto \exp(-\tfrac{1}{2}(n-2) z^2)\, dz$$

approximately. Consequently the distribution of z is approximately normal, with variance $1/(n-2)$. As Fisher has shown, however, a better approximation to the variance of z is $1/(n-3)$.

Example 2. In a random sample of 28 pairs of values from a bivariate normal population, the correlation was found to be 0·7. Is this value consistent with the assumption that the correlation in the population is 0·5?

Here $r = 0\cdot7$, $\rho = 0\cdot5$ and $n = 28$. From the table we find $z = 0\cdot87$ and $\zeta = 0\cdot55$, so that
$$z - \zeta = 0\cdot32.$$

The s.e. of z is $1/\sqrt{(25)} = \tfrac{1}{5}$, so that

$$(z - \zeta)/(\text{s.e.}) = 1\cdot6.$$

Since $z - \zeta$ is considerably less than twice the s.e., its value is not significant. So far as this test goes, the correlation in the population might very well be 0·5.

The 95 % *fiducial limits* for ρ are found in the usual manner (cf. § 51). The value of ζ must be such that

$$|z - \zeta| < 1\cdot96(\text{s.e.}) = 0\cdot392.$$

Consequently
$$0\cdot87 - 0\cdot392 < \zeta < 0\cdot87 + 0\cdot392$$

$$0\cdot48 < \zeta < 1\cdot26,$$

and therefore from the table
$$0\cdot446 < \rho < 0\cdot851.$$

The 95 % fiducial limits for ρ are therefore 0·45 and 0·85 approximately.

95. Comparison of correlations in independent samples

Next suppose that two independent samples of n_1 and n_2 pairs give correlation coefficients of r_1 and r_2 respectively. May they be regarded as drawn from the same population; or is the difference

between r_1 and r_2 significant? On the assumption that the samples
are from the same normal population, the difference between the
values of z for the two samples is normally distributed with S.E.

$$\epsilon = \sqrt{\left(\frac{1}{n_1 - 3} + \frac{1}{n_2 - 3}\right)}.$$

The two values of z are

$$z_1 = \tfrac{1}{2}\log\frac{1 + r_1}{1 - r_1}, \quad z_2 = \tfrac{1}{2}\log\frac{1 + r_2}{1 - r_2}.$$

If $|z_1 - z_2|$ is less than 2ϵ, the difference is not significant at the
5 % level; and the assumption that the samples are from the same
population, or from equally correlated normal populations, is not
discredited.

Example. The first of two samples consists of 23 pairs, and gives a correlation of 0·5; while the second, of 28 pairs, has a correlation of 0·8. Are these
values significantly different?

On the hypothesis that they are from the same normal population, the
S.E. of the difference of the z's is

$$\epsilon = \sqrt{(\tfrac{1}{20} + \tfrac{1}{25})} = \sqrt{0\cdot09} = 0\cdot3.$$

From the table we find that $z_1 = 0\cdot55$ and $z_2 = 1\cdot10$, so that

$$|z_1 - z_2| = 0\cdot55 = 1\cdot83\epsilon,$$

which is a little less than 2ϵ, and is therefore not quite significant at the 5 %
level. The hypothesis is not discredited.

If in the above example the value of r_1 is not given, we may find the limits
between which it must lie in order that $|r_1 - r_2|$ should not be significant at
the 5 % level. For the condition to be satisfied is

$$|z_1 - z_2| < 1\cdot96\epsilon = 0\cdot588.$$

Consequently $$1\cdot10 - 0\cdot588 < z_1 < 1\cdot10 + 0\cdot588$$

$$0\cdot512 < z_1 < 1\cdot688,$$

so that $$0\cdot47 < r_1 < 0\cdot93.$$

96. Combination of estimates of a correlation coefficient*

Suppose that k samples of n_i pairs of values $(i = 1, ..., k)$ yield
correlation coefficients r_i. We may wish to enquire whether the
samples may be regarded as drawn from the same normal population

* Cf. Yates, 1934, 5.

(or equally correlated ones); and, if they may be so regarded, to obtain a combined estimate of the population value ρ. To test the homogeneity of the estimates r_i, we make the assumption that the samples are from equally correlated populations. Then, by means of Fisher's transformation (16), we obtain values z_i of variates which are approximately normally distributed about a common mean, with variances $1/(n_i - 3)$. The estimate of their common mean, ζ, which has minimum variance, is obtained by weighting the values z_i inversely* as their variances. This estimate $\bar{z}$ is therefore

$$\bar{z} = \frac{\sum (n_i - 3) z_i}{\sum (n_i - 3)}. \tag{17}$$

Then, since the variates z_i are approximately normally distributed about ζ, with variances $1/(n_i - 3)$, the sum $\sum (n_i - 3) (z_i - \bar{z})^2$ is distributed approximately as χ^2, with $k - 1$ D.F., the mean $\bar{z}$ having been determined from the data.† The significance of the calculated value of this quantity may be ascertained from the table of χ^2. We may express the above sum in a form more convenient for numerical calculation. Thus

$$\sum_i (n_i - 3) (z_i - \bar{z})^2 = \sum (n_i - 3) z_i^2 - \bar{z}^2 \sum (n_i - 3)$$
$$= \sum (n_i - 3) z_i^2 - [\sum (n_i - 3) z_i]^2 / \sum (n_i - 3).$$

If the calculated value of this expression is not significant as a value of χ^2 with $k - 1$ D.F., the estimates r_i of the correlation in the population may be regarded as homogeneous. In that case the value of $\bar{z}$ given by (17) is an estimate of the true value, ζ, corresponding to the population coefficient ρ. The required estimate of ρ is then given by

$$\rho = \tanh \bar{z},$$

and its value may be read off from the table.

Example. Independent samples of 21, 30, 39, 26 and 35 pairs of values yielded correlation coefficients 0·39, 0·61, 0·43, 0·54 and 0·48 respectively. May these estimates be regarded as homogeneous? If so, find an estimate of the correlation in the population.

* Cf. Ex. IV, 5.
† For details of this part of the proof see Ex. X, 10 below.

The corresponding values of z_i may be taken from the table, and the calculation tabulated as follows:

r_i	z_i	$n_i - 3$	$(n_i - 3)\, z_i$	$(n_i - 3)\, z_i{}^2$
0·39	0·412	18	7·416	3·055
0·61	0·709	27	19·143	13·572
0·43	0·460	36	16·560	7·618
0·54	0·604	23	13·892	8·391
0·48	0·524	32	16·768	8·786
Totals	—	136	73·779	41·422

The value of χ^2 from the data is therefore

$$\chi^2 = 41{\cdot}422 - (73{\cdot}779)^2/136 = 1{\cdot}4 \text{ approximately.}$$

For 4 D.F. this value is not at all significant, so that the coefficients r_i may be regarded as homogeneous. From (17) we have

$$\bar{z} = 73{\cdot}779/136 = 0{\cdot}5425.$$

The table then gives the estimate of ρ for the population as $0{\cdot}495 = 0{\cdot}5$ nearly.

COLLATERAL READING

FISHER, 1938, 2, chapters V, VI and VII.
FISHER, 1915, 2; 1921, 1; 1922, 2; 1924, 2; 1925, 1 and 1935, 1.
TIPPETT, 1931, 2, chapter V.
YULE and KENDALL, 1937, 1, chapter XXIII.
RIDER, 1930, 1; 1939, 6, chapters VI and VIII.
KENNEY, 1939, 3, part II, chapter VII and pp. 172–86.
'STUDENT', 1908, 1 and 2.
SMITH, 1939, 7.
AITKEN, 1939, 1, chapter VII.
MILLS, 1938, 1, chapter XVIII, pp. 598–618.
RIETZ, 1938, 6.
SAWKINS, 1940, 3; 1941, 1.
PITMAN, 1937, 3.
KENDALL, 1943, 2, chapter X and pp. 336–47.

EXAMPLES X

1. A random sample of 16 values from a normal population showed a mean of 41·5 in., and a sum of squares of deviations from this mean equal to 135 in.² Show that the assumption of a mean of 43·5 in. for the population is not reasonable, and that the 95 % fiducial limits for this mean are 39·9 and 43·1 in.

Another sample of 20 values from an unknown population has a mean of 43·0 in., and a sum of squares of deviations from this mean equal to 171 in.^{2} Show that the two samples may be regarded as from the same normal population.

2. Nine patients, to whom a certain drink was administered, registered the following increments in blood pressure: 7, 3, −1, 4, −3, 5, 6, −4, 1. Show that the data do not indicate that the drink was responsible for these increments.

On the null hypothesis the above values are regarded as a random sample from a normal population whose mean is zero. The data give $\bar{x} = 2$ and $s^2 = 63/4$. Hence $t = 1·51$ with $\nu = 8$. This value is not significant.

3. In testing the superiority of Leake's drill over the ordinary drill, plots in the form of long strips were cultivated, two adjacent strips being allotted at random to Leake's drill and the ordinary.* For ten such pairs of plots the values of the excess of the weight of grain from the plot treated by Leake's drill over that obtained by use of the ordinary drill were 2·4, 1·0, 0·7, 0·0, 1·1, 1·6, 1·1, −0·4, 0·1 and 0·7. Show that the data furnish strong evidence of the superiority of Leake's drill.

From the data $\bar{x} = 0·83$ and $\sum (x_i - \bar{x})^2 = 6·001$, so that

$$s^2 = 6·001/9 = 0·666.$$

On the null hypothesis the mean of the population is zero, and $t = 3·22$ for 9 D.F. This value belongs to about the 1 % level of significance. There is thus strong evidence against the null hypothesis.

4. For a random sample of 10 pigs, fed on diet A, the increases in weight in a certain period were

$$10, 6, 16, 17, 13, 12, 8, 14, 15, 9 \text{ lb}.$$

For another random sample of 12 pigs, fed on diet B, the increases in the same period were

$$7, 13, 22, 15, 12, 14, 18, 8, 21, 23, 10, 17 \text{ lb}.$$

* Data from Wishart, 1934, 3, p. 32.

Show that, by the test of § 88, the mean increases of 12 and 15 lb. in the two samples are not significantly different.

$$[s^2 = (120+314)/20 = 21 \cdot 7; \quad t = 1 \cdot 5, \quad \nu = 20.]$$

5. Show that the estimates of the population variance from the samples in Ex. 4 are not significantly different.

6. Deduce from § 82 (25), that the statistic $r \sqrt{(n-2)}/\sqrt{(1-r^2)}$ conforms to the t distribution for $n - 2$ D.F.

7. A random sample of 18 pairs from a bivariate normal population showed a correlation coefficient of 0·3. Is this value significant of correlation in the population? Prove by the method of § 89, Ex. 2, that the least value of r significant at the 5 % level is about 0·47.

8. A random sample of 19 pairs from a bivariate normal population showed a correlation of 0·65. Prove that this is consistent with the assumption of a correlation of 0·40 in the population. Also show that the 95 % fiducial limits for ρ are 0·28 and 0·85 approximately.

A second sample, of 23 pairs, showed a correlation of 0·40. Prove by the method of § 95 that the two samples may be regarded as from equally correlated populations.

9. Show that the mean value of the positive square root of a Beta variate of the second kind, with parameters l and m, is

$$\Gamma(l + \tfrac{1}{2})\,\Gamma(m - \tfrac{1}{2})/\Gamma(l)\,\Gamma(m).$$

Deduce that the mean value of $|t|$ for ν D.F. is

$$\Gamma(\tfrac{1}{2}(\nu - 1))\,\sqrt{\nu}/\Gamma(\tfrac{1}{2}\nu)\,\sqrt{\pi}.$$

Also show that, for samples from an uncorrelated bivariate normal population in which the variances are σ_1^2 and σ_2^2, the mean value of the modulus of the regression coefficient of y on x is

$$\sigma_2\,\Gamma(\tfrac{1}{2}(n-2))/\sigma_1\,\Gamma(\tfrac{1}{2}(n-1))\,\sqrt{\pi}.$$

10. To prove that the sum $\sum\limits_{i} (n_i - 3)(z_i - \bar{z})^2$ of § 96 is distributed approximately as χ^2 with $k-1$ D.F., we may proceed by the method of § 77. Denoting the sum by T^2 we have

$$
\begin{aligned}
T^2 &= \sum (n_i - 3)\left[(z_i - \zeta) - (\bar{z} - \zeta)\right]^2 \\
&= \sum (n_i - 3)(z_i - \zeta)^2 - 2(\bar{z} - \zeta)\sum (n_i - 3)(z_i - \zeta) + (\bar{z} - \zeta)^2 \sum (n_i - 3) \\
&= \sum (n_i - 3)(z_i - \zeta)^2 - \xi_1^2,
\end{aligned} \tag{i}
$$

where $\qquad \xi_1 = \sum (n_i - 3)(z_i - \zeta)/\sqrt{\sum (n_i - 3)}.$

Now introduce an orthogonal linear transformation

$$
\xi_i = \sum_j c_{ij} x_j \quad (j = 1, 2, \ldots, k),
$$

where $\qquad x_j = (z_j - \zeta)\sqrt{(n_j - 3)},$

and the first of the ξ's is identical with ξ_1 above. Then, since the x's are independently and normally distributed about zero with unit S.D., so are the ξ's; and in virtue of (i)

$$
T^2 = \sum_{i=1}^{k} x_i^2 - \xi_1^2 = \sum_{i=1}^{k} \xi_i^2 - \xi_1^2 = \sum_{i=2}^{k} \xi_i^2.
$$

Consequently T^2 is distributed like χ^2 with $k-1$ D.F.

11. Show that, for the t distribution with ν D.F., the moment of order r (even) about the origin is, for $r < \nu$,

$$
(r-1)(r-3)\ldots 1 \cdot \nu^{\frac{1}{2}r}/(\nu - 2)(\nu - 4)\ldots(\nu - r).
$$

12. For samples of n from a population with the exponential distribution $f(x) = e^{-x}$, $x \geqslant 0$, show that the range conforms to

$$
\phi(w) = (n-1)\,e^{-w}(1 - e^{-w})^{n-2},
$$

and hence that $\quad E(w) = 1 + \tfrac{1}{2} + \tfrac{1}{3} + \tfrac{1}{4} + \ldots + \dfrac{1}{n-1}.$

CHAPTER XI

ANALYSIS OF VARIANCE AND COVARIANCE

ANALYSIS OF VARIANCE

97. Resolution of the 'sum of squares'

In the words of its author, R. A. Fisher, analysis of variance is the 'separation of the variance ascribable to one group of causes from the variance ascribable to other groups'.* It is a procedure by which the variation embodied in the data of the sample may be resolved into component variations due to independent factors. Each of the components yields an estimate of the population variance; and these estimates are tested for homogeneity by means of the F table.

Consider a random sample of N values of a normally distributed variable x. It is frequently possible to arrange these in classes according to a certain factor or criterion. For instance, if the variable is the price of a certain commodity, the classes may correspond to different seasons or to different districts. Or, if the variable is the crop yield of a variety of cereal, the classes may correspond to different manurial treatments. Let x_{ij} denote the value of the jth member in the ith class. Thus the first subscript indicates the class, and the second the position in that class. Let n_i be the number of members in the ith class, $\bar{x}_i$ the mean value for that class, and $\bar{x}$ the general mean for the whole sample of N values.

Then
$$N\bar{x} = \sum_i \sum_j x_{ij}, \quad \sum_i \sum_j (x_{ij} - \bar{x}) = 0 \qquad (1)$$

and
$$n_i \bar{x}_i = \sum_j x_{ij}, \quad \sum_j (x_{ij} - \bar{x}_i) = 0. \qquad (2)$$

To resolve the sum of squares of the deviations of the N values x_{ij} from the general mean, we may do so first for the members of the ith class. Thus, in virtue of § 3 (10),

$$\sum_j (x_{ij} - \bar{x})^2 = \sum_j (x_{ij} - \bar{x}_i)^2 + n_i (\bar{x}_i - \bar{x})^2,$$

* Fisher, 1938, 2, p. 216.

H

and on summing this result for all the classes we have the required resolution of the 'sum of squares'

$$\sum_i \sum_j (x_{ij} - \bar{x})^2 = \sum_i \sum_j (x_{ij} - \bar{x}_i)^2 + \sum_i n_i(\bar{x}_i - \bar{x})^2. \qquad (3)$$

This formula holds, of course, whether the population is normal or not.

98. Homogeneous population. One criterion of classification

Now suppose that the population, from which the random sample of N values was drawn, is homogeneous with respect to the factor of classification, that is to say, that the factor has no effect upon the value of the variate. Then if the population is divided into classes according to this factor, the different classes will have the same statistical properties. In particular, they will have the same mean μ and the same variance σ^2, which are the mean and the variance of the population. Then, from the various sums in (3), we can obtain three unbiased estimates of σ^2. For, by (7) and (8) of §54, the expected value of the first sum in sampling is $(N-1)\sigma^2$; so that this sum, divided by $N-1$, gives an unbiased estimate of σ^2 based on $N-1$ D.F. Similarly, the n_i values in the ith class constitute a random sample whose mean is $\bar{x}_i$, so that

$$E[\sum_j (x_{ij} - \bar{x}_i)^2] = (n_i - 1)\sigma^2,$$

and therefore*

$$E[\sum_i \sum_j (x_{ij} - \bar{x}_i)^2] = \sum_i (n_i - 1)\sigma^2 = (N-h)\sigma^2,$$

where h is the number of classes. Thus the second sum in (3), divided by $N-h$, gives an unbiased estimate of σ^2 based on $N-h$ D.F. And, since the expected values of the two members of (3) must be equal, that of the final sum* is $(h-1)\sigma^2$; so that this sum, divided by $h-1$, gives an unbiased estimate of σ^2 based on $h-1$ D.F. The identity

$$(N-1)\sigma^2 = (N-h)\sigma^2 + (h-1)\sigma^2 \qquad (4)$$

obtained by taking expected values of the various sums in (3), shows that degrees of freedom are additive, the number of freedoms

* See also Ex. XI, 1, below.

corresponding to the total sum being equal to the sum of the freedoms corresponding to the partial sums. The above results are usually tabulated as follows:

Source of variation	D.F.	Sum of squares	Mean square
Between class means	$h-1$	$\sum_i n_i(\bar{x}_i - \bar{x})^2$	$\sum_i n_i(\bar{x}_i - \bar{x})^2/(h-1)$
Within classes	$N-h$	$\sum_i \sum_j (x_{ij} - \bar{x}_i)^2$	$\sum_i \sum_j (x_{ij} - \bar{x}_i)^2/(N-h)$
Total	$N-1$	$\sum_i \sum_j (x_{ij} - \bar{x})^2$	—

In the columns headed 'D.F.' and 'Sum of squares' the items are additive, but not in the last column which gives the estimates of σ^2.

The argument so far holds whether the population is normal or not. In the case of a homogeneous *normal population* the results follow from the distributions of the various sums in (3). For then the first sum divided by σ^2 is distributed like χ^2 with $N-1$ D.F., as proved in §77. The mean value of this sum is therefore $(N-1)\sigma^2$. Similarly, $\sum_j (x_{ij} - \bar{x}_i)^2/\sigma^2$ is a χ^2 with $n_i - 1$ D.F.; and therefore the second sum in (3), divided by σ^2, is distributed like χ^2 with

$$\sum_i (n_i - 1) = (N-h) \text{ D.F.}$$

The mean value of this sum is therefore $(N-h)\sigma^2$. Similarly for the final sum in (3).

In order to *test the homogeneity* of the estimates of σ^2 by means of the variance ratio and the F table, it is necessary to assume that the population is normal; for this test is founded on that assumption. The practice is to compare the estimate 'between class means' with that obtained 'within classes'. For the final sum in (3) represents the variation due to the factor of classification, and the second sum in (3) is the residual variation after the former has been removed. If the estimate obtained between classes is significantly greater than that within classes, we are justified in concluding that the factor of classification exercises an influence on the value of the variable. In that case the assumption of homogeneity is discredited, and we must regard the population as heterogeneous. If, however, the estimates of σ^2 are not significantly different, the test provides no evidence against the hypothesis of a homogeneous population. It is important to remember that the variance ratio may be tested by means of the F table only if the two estimates of variance are statistically independent. Since the mean of a random sample from a normal population is distributed independently of its variance, the two sums in the second member of (3) are independent, and the required condition is satisfied.*

99. Calculation of the sums of squares

In calculating the above sums of squares it is not necessary to find the deviations from the various means. We know that the sum of the squares of the deviations of N numbers x_s $(s = 1, ..., N)$, from their mean $\bar{x}$ is, in virtue of § 3 (11),

$$\sum_s (x_s - \bar{x})^2 = \sum_s x_s^2 - N\bar{x}^2 = \sum_s x_s^2 - T^2/N,$$

where T is the sum of the numbers. Applying this formula to the various sums considered above we have

$$\sum_i \sum_j (x_{ij} - \bar{x})^2 = \sum_i \sum_j x_{ij}^2 - T^2/N, \tag{5}$$

the grand total T being given by

$$T = \sum_i \sum_j x_{ij}.$$

Similarly, summing first for the values in the ith class, we have

$$\sum_i \sum_j (x_{ij} - \bar{x}_i)^2 = \sum_i (\sum_j x_{ij}^2 - T_i^2/n_i),$$

where T_i is the sum of the values in the ith class. Consequently

$$\sum_i \sum_j (x_{ij} - \bar{x}_i)^2 = \sum_i \sum_j x_{ij}^2 - \sum_i T_i^2/n_i. \tag{6}$$

Subtracting (6) from (5) we find for the third sum of squares

$$\sum_i n_i(\bar{x}_i - \bar{x})^2 = \sum_i T_i^2/n_i - T^2/N. \tag{7}$$

---

* See also Fisher, 1925, 1 and Irwin, 1934, 4.

This result may also be obtained directly. For, since n_i is the frequency of the value $\bar{x}_i$,

$$\sum_i n_i(\bar{x}_i - \bar{x})^2 = \sum_i n_i \bar{x}_i^2 - (\sum_i n_i \bar{x}_i)^2/N = \sum_i T_i^2/n_i - T^2/N$$

as stated.

Since the deviations from the means are independent of the choice of origin, the results obtained by using (5), (6) and (7) are unaltered by a change of origin. In other words, if all the values x_{ij} are decreased (or increased) by the same constant, the values obtained for the three sums of squares are unchanged. The arithmetic may often be simplified in this way, large numbers being replaced by much smaller ones.

Ex. 35 plots, of approximately equal fertility, were sown with 7 different varieties of wheat, 5 plots to each variety, the distribution of varieties among the plots being random. The following table gives the yields of grain in bushels per acre, the 7 columns corresponding to the different varieties. Do the data (fictitious) indicate a significant difference in the yields of the varieties?

13	15	14	14	17	15	16
11	11	10	10	15	9	12
10	13	12	15	14	13	13
16	18	13	17	19	14	15
12	12	11	10	12	10	11

The classification is according to variety. The number of classes is $h = 7$, and the number of items in each class is $n_i = 5$. Consequently $N = 35$. The arithmetic is simplified by shifting the origin to (say) $x = 12$. Diminishing all the yields by 12 we may rewrite the table:

1	3	2	2	5	3	4
-1	-1	-2	-2	3	-3	0
-2	1	0	3	2	1	1
4	6	1	5	7	2	3
0	0	-1	-2	0	-2	-1

from which we have

$$T_i = 2,\ 9,\ 0,\ 6,\ 17,\ 1,\ 7; \quad T = 42;$$

$$\bar{x}_i = 0{\cdot}4,\ 1{\cdot}8,\ 0,\ 1{\cdot}2,\ 3{\cdot}4,\ 0{\cdot}2,\ 1{\cdot}4.$$

Hence $\quad \sum_i \sum_j x_{ij}^2 = 266, \quad T^2/N = 50{\cdot}4, \quad \sum_i T_i^2/n_i = 92.$

The three sums of squares are therefore, by (5), (6) and (7),

$$\sum_i \sum_j (x_{ij} - \bar{x})^2 = 266 - 50{\cdot}4 = 215{\cdot}6,$$

$$\sum_i \sum_j (x_{ij} - \bar{x}_i)^2 = 266 - 92 = 174,$$

$$\sum_i n_i (\bar{x}_i - \bar{x})^2 = 92 - 50{\cdot}4 = 41{\cdot}6.$$

In tabular form:

Source of variation	D.F.	Sum of squares	Mean square	F
Between varieties	6	41·6	6·933	1·1
Within varieties	28	174	6·214	—
Total	34	215·6	—	—

For $\nu_1 = 6$ and $\nu_2 = 28$ the value $1{\cdot}1$ of F is not significant. Since the estimates of variance between varieties and within varieties are not significantly different, the experiment as a whole does not indicate significant variation in the yields of varieties.

100. Two criteria of classification

Consider next the case in which the N values x_{ij} of the data may be classified according to two different criteria, A and B. For simplicity suppose that A determines h different classes, and B determines k different groups; also that the hk values of the variable are such that, in each of the h classes there is one value from each group, and in each of the k groups one value from each class. For the purpose of calculation the hk values may be arranged in a rectangular array of h columns and k rows, the columns corresponding to classification A and the rows to B. The double suffix notation will indicate that x_{ij} belongs to the ith class and the jth group; and, in the rectangular array, this value occurs in the ith column and the jth row. As before $\bar{x}$ denotes the general mean; $\bar{x}_i$ is the mean of the values in the ith class, and $\bar{x}_j$ the mean of those in the jth group.

The argument leading to (3) is valid here also, the resolution expressed by that equation being with respect to the means of columns. We may use (3) again to resolve the sum $\sum_i \sum_j (x_{ij} - \bar{x}_i)^2$, this time, however, with respect to the means of rows. In doing so we take as a new variable

$$X_{ij} = x_{ij} - \bar{x}_i,$$

each value of the original variable being diminished by the mean of the column in which it lies. The values X_{ij} may be arranged in h columns and k rows corresponding to those of x_{ij}. Now the general mean of X_{ij} is zero, since the means of x_{ij} and $\bar{x}_i$ are each $\bar{x}$. Thus $\bar{X} = 0$. Similarly the mean of the values X_{ij} in the jth row is

$$\bar{X}_j = \bar{x}_j - \bar{x}.$$

Accordingly, on applying (3) to the quantities X_{ij}, but taking row means instead of column means, we have

$$\sum_i \sum_j (X_{ij} - \bar{X})^2 = \sum_j h(\bar{X}_j - \bar{X})^2 + \sum_i \sum_j (X_{ij} - \bar{X}_j)^2,$$

or its equivalent

$$\sum_i \sum_j (x_{ij} - \bar{x}_i)^2 = \sum_j h(\bar{x}_j - \bar{x})^2 + \sum_i \sum_j (x_{ij} - \bar{x}_i - \bar{x}_j + \bar{x})^2. \tag{8}$$

Substituting this value in (3), and remembering that the numbers n_i are each equal to k, we have the resolution expressed by

$$\sum_i \sum_j (x_{ij} - \bar{x})^2 = \sum_i k(\bar{x}_i - \bar{x})^2 + \sum_j h(\bar{x}_j - \bar{x})^2 + \sum_i \sum_j Y_{ij}^2, \tag{9}$$

where $$Y_{ij} = x_{ij} - \bar{x}_i - \bar{x}_j + \bar{x}.$$

As in the preceding section, the expected value of the first member of (9) is $(hk - 1)\sigma^2$. Similarly, on the assumption of a homogeneous population, the expected values of the first two sums in the second member are $(h - 1)\sigma^2$ and $(k - 1)\sigma^2$ respectively. That of the final sum is therefore $(hk - h - k + 1)\sigma^2$. Thus the various sums in (9), divided by $(hk - 1)$, $(h - 1)$, $(k - 1)$ and $(h - 1)(k - 1)$ respectively, give unbiased estimates of the population variance based on degrees of freedom represented by these divisors. When the population is *normal* the four sums in (9), divided by σ^2, are distributed like χ^2

with degrees of freedom as stated above; and the mean values of the various sums follow from this. The above results are usually tabulated:

Variation	D.F.	Sum of squares	Mean square
Between classes	$h-1$	$\sum_i k(\bar{x}_i - \bar{x})^2$	
Between groups	$k-1$	$\sum_j h(\bar{x}_j - \bar{x})^2$	The quotient of 'sum of squares' by D.F. in each case
Error	$(h-1)(k-1)$	$\sum_i \sum_j Y_{ij}^2$	
Total	$hk-1$	$\sum_i \sum_j (x_{ij} - \bar{x})^2$	—

Degrees of freedom are additive, as well as sums of squares. The variation corresponding to the factors of classification is represented by the first two sums in the second member of (9). The uncontrolled variation represented by the residual sum $\sum_i \sum_j Y_{ij}^2$ is due to a variety of causes, which are grouped under the term 'error'.

To test the hypothesis of homogeneity in the population, we compare the two estimates of variance obtained between classes and between groups with that obtained from error. If the factor corresponding to either classification has a significant effect upon the value of the variable, this will appear in the corresponding mean square. In order to test the variance ratio by means of the F table* the population must be assumed normal. If either of the first two estimates of variance is significantly different from that obtained from error, the hypothesis of homogeneity is discredited. The significance of the difference of the means of any two classes, or any two groups, may be tested by means of the t table, as in the example below.

Example. An agricultural experiment was conducted to test the effects of change of soil (5 blocks) and variety of wheat (7 different strains) on the yield of grain. Each block was divided into seven plots, and the plots of each block were assigned at random to the seven varieties. The yields, in bushels per acre, are set out in the same rectangular array as in the example of § 99, columns corresponding to varieties and rows to blocks. Discuss the significance of the variation of yield with the two factors.

* The independence of these estimates of variance may be established by Cochran's method. See 1934, 6.

With origin at $x = 12$ the values of T, T_i, $\bar{x}_i$, the total sum of squares and the sum of squares corresponding to varieties are as already found. Show similarly that

$$T_j = 20, \qquad -6, \qquad 6, \qquad 28, \qquad -6$$

$$\bar{x}_j = 2\cdot86, \qquad -0\cdot86, \quad 0\cdot86, \qquad 4, \qquad -0\cdot86$$

$$\sum_j T_j^2/7 = 184\cdot6, \quad \sum_j T_j^2/7 - T^2/35 = 134\cdot2.$$

The tabulation of results is:

Source of variation	D.F.	Sum of squares	Mean square	F
Varieties	6	41·6	6·93	4·2**
Blocks	4	134·2	33·55	20·2**
Error	24	39·8	1·66	––
Total	34	215·6	—	—

The double asterisk indicates that the value of F is significant at the 1 % level (a single asterisk denoting significance at the 5 % level). Thus the yields of the varieties are significantly different; and the experiment indicates a very marked variation in soil fertility from one block to another.

We may use the t test to examine more closely the difference between the mean yields of any two varieties, say the second and third. The difference is $\bar{x}_2 - \bar{x}_3 = 1\cdot8$. To test the significance of this we use the estimate of variance obtained from 'error', and corresponding to 24 D.F. This value, 1·66, is the estimated variance of the yield of a single plot. The estimated variance of the mean of 5 plots is thus 1·66/5; and that of the difference of the means of two independent samples of 5 plots each is $1\cdot66 \times 2/5 = 0\cdot664$. The s.e. of the difference of the means of two varieties is therefore 0·815 nearly. The value of t for the above two varieties is thus $1\cdot8/0\cdot815 = 2\cdot2$. For 24 D.F. this is significant at the 5 % level. The least difference, m, that is significant is given by $m/0\cdot815 = 2\cdot06$, so that $m = 1\cdot68$ nearly. Show also that the s.e. of the difference between the mean yields for two blocks is 0·69, and that a difference of 1·4 is significant at the same level.

101. The Latin square. Three criteria of classification

In the general case of three criteria of classification, corresponding to h, k, p classes respectively, we should require a three-dimensional generalization of the rectangular array employed above. In the particular case for which $h = k = p$ this requirement is obviated by an arrangement known as the Latin square. As it is customary to use the letters A, B, C, ... to distinguish the different classes of one of the three classifications, we may conveniently explain the Latin square, of order n, as an arrangement of the n letters A, B, C, ... in the form of a square array, such that each letter occurs once in each column and once in each row. Consequently each letter occurs n times in a Latin square of order n. The accompanying array is one arrangement for a Latin square of order five.

$$
\begin{array}{ccccc}
B & D & E & A & C \\
C & A & B & E & D \\
D & C & A & B & E \\
E & B & C & D & A \\
A & E & D & C & B
\end{array}
$$

The triple classification and the Latin square arrangement are illustrated by the design of the following agricultural experiment. The variable is the yield of grain per acre, and the object of the experiment is to test the effect on yield due to change of manurial treatment, and to variation of soil in each of two perpendicular directions. A block of land is divided into n^2 plots, arranged in n parallel rows in one of the given directions and n parallel columns in the perpendicular direction. The n different treatments are distributed at random among the n plots of each row, but in such a way that no two plots in the same column are given the same treatment. We thus have a Latin square in which letters correspond to treatments, while rows and columns correspond to soil variation in the given perpendicular directions.

The necessary formulae for calculation are obtained by an easy extension of the foregoing results. With the same notation as in the preceding sections, the resolution expressed by (9) is still valid. The final sum $\sum_i \sum_j Y_{ij}^2$ may be separated into components by applying (3) to the variable Y_{ij} and resolving, not with respect to the means of rows or columns, but with respect to the means of letters. Let $\overline{Y}$ denote the general mean of the values Y_{ij}, and $\overline{Y}_l$ the mean of the values for the lth letter. Then, since the mean of each of the four terms in Y_{ij} has the same magnitude $\bar{x}$, we have $\overline{Y} = 0$. To calculate $\overline{Y}_l$ we consider the four terms of Y_{ij} separately. The mean value of

x_{ij} corresponding to the lth letter we shall denote by $\bar{x}_l$. The mean of the values $\bar{x}_i$ for the lth letter is $\bar{x}$, since each column contains the lth letter once only. Similarly the mean of the values $\bar{x}_j$ for the lth letter is $\bar{x}$. The last term of Y_{ij} is constant, and we have the result

$$\bar{Y}_l = \bar{x}_l - \bar{x} - \bar{x} + \bar{x} = \bar{x}_l - \bar{x}.$$

Thus, on applying (3) to the values Y_{ij} and the letters, we deduce

$$\sum_i \sum_j (Y_{ij} - \bar{Y})^2 = n \sum_l (\bar{Y}_l - \bar{Y})^2 + \sum_i \sum_j (Y_{ij} - \bar{Y}_l)^2$$

or its equivalent

$$\sum_i \sum_j Y_{ij}^2 = n \sum_l (\bar{x}_l - \bar{x})^2 + \sum_i \sum_j (x_{ij} - \bar{x}_i - \bar{x}_j - \bar{x}_l + 2\bar{x})^2.$$

Substituting this value in (9), and putting $h = k = n$, we obtain the required resolution

$$\sum_i \sum_j (x_{ij} - \bar{x})^2 = n \sum_i (\bar{x}_i - \bar{x})^2 + n \sum_j (\bar{x}_j - \bar{x})^2 + n \sum_l (\bar{x}_l - \bar{x})^2 + \sum_i \sum_j Z_{ij}^2,$$

(10)

where $$Z_{ij} = x_{ij} - \bar{x}_i - \bar{x}_j - \bar{x}_l + 2\bar{x}.$$

The expected value of the first member of (10) is $(n^2 - 1)\sigma^2$, and on the assumption of a homogeneous population, that of each of the first three sums on the right is $(n-1)\sigma^2$, as proved in §98. Consequently

$$E\left(\sum_i \sum_j Z_{ij}^2\right) = [n^2 - 1 - 3(n-1)]\sigma^2 = (n-1)(n-2)\sigma^2.$$

Thus each of the first three sums in the second member of (10), divided by $(n-1)$, gives an unbiased estimate of σ^2 based on $(n-1)$ D.F.; and the final sum, divided by $(n-1)(n-2)$, gives an unbiased estimate based on that number of freedoms. In the case of a normal population the various sums in (10), divided by σ^2, are distributed like χ^2 with D.F. equal to those of the corresponding estimates of σ^2. The results may be tabulated as shown below. The first three mean squares are compared with that obtained from error in the usual manner. The significance of the difference of the means of two classes may be tested by the t table as illustrated earlier.

The various sums of squares may be calculated from the usual formulae, putting $N = n^2$ and $h = k = n$. Thus

$$\sum_i \sum_j (x_{ij} - \bar{x})^2 = \sum_i \sum_j x_{ij}^2 - T^2/n^2.$$

Similarly,

$$n \sum_l (\bar{x}_l - \bar{x})^2 = \sum_l T_l^2/n - T^2/n^2,$$

where T_l is the sum of the values for the lth letter; and so on. By subtraction we have therefore

$$\sum_i \sum_j Z_{ij}^2 = \sum_i \sum_j x_{ij}^2 - \left(\sum_i T_i^2 + \sum_j T_j^2 + \sum_l T_l^2\right)/n + 2T^2/n^2.$$

It is usual to find this sum by subtraction after the other sums have been calculated.

Source of variation	D.F.	Sum of squares	Mean square
Columns	$n-1$	$n\sum_i (\bar{x}_i - \bar{x})^2$	
Rows	$n-1$	$n\sum_j (\bar{x}_j - \bar{x})^2$	The quotient of 'sum of squares' by D.F. in each case
Treatments	$n-1$	$n\sum_l (\bar{x}_l - \bar{x})^2$	
Error	$(n-1)(n-2)$	$\sum_i \sum_j Z_{ij}^2$	
Total	n^2-1	$\sum_i \sum_j (x_{ij} - \bar{x})^2$	—

Example. An agricultural experiment was conducted on the Latin square plan to test the effect on yield due to change of treatment (5 kinds) and also to variation of soil in each of two perpendicular directions. The results are set out in the Latin square below ($n = 5$), in which letters correspond to treatments, while rows and columns correspond to the two perpendicular directions. Are the effects on yield significant?

A	7·4	D	8·9	E	5·8	B	12·0	C	14·3
C	11·8	B	6·5	A	8·7	E	7·6	D	7·9
D	10·1	C	17·9	B	9·0	A	8·5	E	7·1
E	8·8	A	10·1	C	15·7	D	11·1	B	7·4
B	11·8	E	8·8	D	14·3	C	18·4	A	10·1

Shifting the origin to $x = 10$ (i.e. reducing each of the above yields by 10) the reader may easily verify that

$$T_i = -0\cdot1, \quad 2\cdot2, \quad 3\cdot5, \quad 7\cdot6, \quad -3\cdot2.$$
$$T_j = -1\cdot6, \quad -7\cdot5, \quad 2\cdot6, \quad 3\cdot1, \quad 13\cdot4; \quad T = 10.$$
$$T = -5\cdot2, \quad -3\cdot3, \quad 28\cdot1, \quad 2\cdot3, \quad -11\cdot9.$$

Consequently $\qquad\qquad T^2/N = 10 \times 10/25 = 4\cdot00.$

The various sums of squares are given by

$$\Sigma\Sigma x_{ij}^2 - T^2/N = 285\cdot18 - 4\cdot00 = 281\cdot18 \quad \text{(Total)}$$

$$\underset{i}{\Sigma} T_i^2/n - T^2/N = 17\cdot02 - 4\cdot00 = 13\cdot02 \quad \text{(Columns)}$$

$$\underset{j}{\Sigma} T_j^2/n - T^2/N = 50\cdot95 - 4\cdot00 = 46\cdot95 \quad \text{(Rows)}$$

$$\underset{l}{\Sigma} T_l^2/n - T^2/N = 194\cdot89 - 4\cdot00 = 190\cdot89 \quad \text{(Treatments)}$$

$$\text{Remainder} = 30\cdot32 \quad \text{(Error)}$$

The tabulation is:

Source of variation	D.F.	Sum of squares	Mean square	F
Columns	4	13·02	3·255	1·3
Rows	4	46·95	11·737	4·6*
Treatments	4	190·89	47·722	18·9**
Error	12	30·32	2·527	—
Total	24	281·18	—	—

For $\nu_1 = 4$ and $\nu_2 = 12$ the 5 and 1 % values of F are 3·26 and 5·41. Thus the variation in rows is significant, and that due to treatments is highly significant.

Show that the s.e. of the difference of the means of two classes is 1·005, and hence that a difference of means less than 2·2 is not significant at the 5 % level. Thus the yield due to treatment C is significantly greater than that due to any other treatment; and the yield due to D is significantly greater than that due to E.

102. Significance of an observed correlation ratio

We shall next consider some applications of analysis of variance† to testing the significance of an observed correlation ratio, coefficient or index, and to testing the linearity of a regression. First suppose that a value, η, of the correlation ratio of y on x, is obtained from a random sample of N pairs of values from a bivariate population in which y is normally distributed. We wish to test whether this value is significant of an association between the two variables in the population. Let the N pairs of values of the variables be arranged in arrays as in Chapter v, so that the y's are classified

† According to the definition given by Wishart and Sanders (see § 105, below) these applications should be classified under 'Analysis of Covariance'.

according to the corresponding values of x. Then since the subscript i indicates the array, and j the position in that array, we have as in (3)

$$\sum_i \sum_j (y_{ij} - \bar{y})^2 = \sum_i \sum_j (y_{ij} - \bar{y}_i)^2 + \sum_i n_i (\bar{y}_i - \bar{y})^2, \qquad (11)$$

$\bar{y}_i$ being the mean value of y in the ith array, and n_i the number of values in that array. In virtue of §34 this equation corresponds to the identity

$$NS_2^2 = NS_2^2(1 - \eta^2) + NS_2^2 \eta^2, \qquad (12)$$

S_2^2 being the variance of y in the sample.

On the assumption that there is no association between the two variables in the population, the y's in each array may be regarded as a random sample from the population of y's. The various sums in (11), divided by the variance σ^2 of y in the population, are then distributed like χ^2 with $(N-1)$, $(N-h)$ and $(h-1)$ D.F., h being the number of arrays. The problem is therefore the same as in §98. On taking expected values of both members of (11) we have

$$(N-1)\sigma^2 = (N-h)\sigma^2 + (h-1)\sigma^2. \qquad (13)$$

The various sums in (11), divided by the corresponding coefficients in (13), yield unbiased estimates of σ^2. The tabulation is:

Source of variation	D.F.	Sum of squares	Mean square
Between arrays	$h-1$	$NS_2^2\eta^2$	$NS_2^2\eta^2/(h-1)$
Within arrays	$N-h$	$NS_2^2(1-\eta^2)$	$NS_2^2(1-\eta^2)/(N-h)$
Total	$N-1$	NS_2^2	—

To test whether the mean square between arrays is significantly greater than that within arrays we have

$$F = \frac{\eta^2}{1-\eta^2} \times \frac{N-h}{h-1},$$

with
$$\nu_1 = h-1, \quad \nu_2 = N-h.$$

Example. A random sample of 79 pairs, arranged in 7 arrays of y's, gave a correlation ratio of y on x equal to $0 \cdot 4$. Is this value significant?

Here
$$\nu_1 = 6, \quad \nu_2 = 72,$$

$$F = \frac{0 \cdot 16}{0 \cdot 84} \times \frac{72}{6} = \frac{16}{7} = 2 \cdot 29*.$$

Since the 5 % value of F is $2 \cdot 23$, the above value is significant at that level. We conclude that there is association between the variables in the population.

103. Significance of a regression function

To test the significance of a regression function is to examine whether the data of the sample indicate any degree of association of the variables, of the type represented by the regression equation. We shall see that, in the case of a linear regression equation, this amounts to testing the significance of an observed value of r; while, in the case of an equation of curvilinear regression, it is equivalent to testing the significance of an observed value of the correlation index R.

Consider first a linear regression equation for a random sample of N pairs of values from a bivariate normal population. If Y_i is the estimate of y_i given by the regression equation, we know by § 27 (31) that

$$\sum_i \sum_j (y_{ij} - \bar{y})^2 = \sum_i \sum_j (y_{ij} - Y_i)^2 + \sum_i n_i (Y_i - \bar{y})^2. \tag{14}$$

The first sum of squares in the second member is due to deviations from the regression function, and the second sum to the regression function itself. In virtue of § 27 the various sums in (14) are equal to the corresponding terms in the identity

$$NS_2^2 = NS_2^2(1 - r^2) + Nr^2 S_2^2.$$

On the assumption of an uncorrelated normal population these sums, divided by σ^2, are distributed like χ^2 with $N-1$, $N-2$ and 1 D.F. respectively, as proved in § 82. The expected values of the various sums in (14) are therefore the corresponding terms of the identity

$$(N-1)\sigma^2 = (N-2)\sigma^2 + \sigma^2,$$

and each sum thus gives an unbiased estimate of σ^2, on division by the appropriate number of D.F. The tabulated results are:

Source of variation	D.F.	Sum of squares	Mean square
Regression function	1	$\sum_i n_i (Y_i - \bar{y})^2$	$Nr^2 S_2^2$
Deviation from the regression function	$N-2$	$\sum_i \sum_j (y_{ij} - Y_i)^2$	$NS_2^2(1 - r^2)/(N-2)$
Total	$N-1$	$\sum_i \sum_j (y_{ij} - \bar{y})^2$	—

If the mean square due to the regression function is significantly greater than that due to deviations from this function, we conclude that there is a real association between the variables, of the type indicated by the regression equation. To test the significance we have for the variance ratio

$$F = (N-2)\, r^2/(1-r^2); \qquad \nu_1 = 1, \quad \nu_2 = N-2.$$

The method is thus equivalent to testing the significance of r; and it will be noted that the above statistic F is the square of the statistic t of § 89 (8). The two tests are therefore equivalent.

Example. A correlation of 0·5 is obtained from a random sample of 26 pairs from a normal population. Is this value significant?

Here $\quad r = \tfrac{1}{2}, \quad N = 26, \quad F = 8^{**}, \quad \nu_1 = 1, \quad \nu_2 = 24.$

From the table we see that this value of F is highly significant of correlation in the population.

Obtain the same result by use of the t table.

In the case of a correlation index R associated with a curved regression line, the formulae of § 41 show that the above argument holds if r^2 is replaced by R^2, except that the numbers of D.F. must be altered. If k is the number of statistics that must be calculated from the sample to obtain the regression equation, the number of D.F. associated with deviations from the regression function is $N-k$, and the number associated with the regression function itself is $k-1$. For testing the significance of the regression function we have therefore

$$F = \frac{R^2}{1-R^2} \times \frac{N-k}{k-1}; \qquad \nu_1 = k-1, \quad \nu_2 = N-k.$$

104. Test for non-linearity of regression

Considering the random sample of N pairs of values from a bivariate normal population, let us return to equation (11). In virtue of § 36 (18), the final sum in that equation may be further resolved as

$$\sum_i n_i(\overline{y}_i - \overline{y})^2 = \sum_i n_i(\overline{y}_i - Y_i)^2 + \sum_i n_i(Y_i - \overline{y})^2, \qquad (15)$$

the various sums being equal to the corresponding terms of the identity

$$N\eta^2 S_2^2 = N(\eta^2 - r^2)\, S_2^2 + Nr^2 S_2^2.$$

The three sums in (15), divided by σ^2, are distributed like χ^2 with $h-1$, $h-2$ and 1 D.F. respectively; and on taking expected values of both members of (15) we obtain the identity

$$(h-1)\sigma^2 = (h-2)\sigma^2 + \sigma^2.$$

The various sums in (15), divided by the corresponding numbers of D.F., give unbiased estimates of σ^2. That obtained from the first sum on the right is associated with deviations of the means of arrays from the line of regression of y on x. On the assumption of linearity of regression this sum is due to sampling errors; and the estimate of σ^2 obtained from it should not be significantly greater than that derived from the sum of squares within arrays, i.e. from

$$\sum_i \sum_j (y_{ij} - \bar{y}_i)^2.$$

Tabulating as usual we have:

Source of variation	D.F.	Sum of squares	Mean square
Linear regression	1	$\sum_i n_i (Y_i - \bar{y})^2$	$Nr^2 S_2^2$
Deviation of means from regression line	$h-2$	$\sum_i n_i (\bar{y}_i - Y_i)^2$	$N(\eta^2 - r^2) S_2^2/(h-2)$
Within arrays	$N-h$	$\sum_i \sum_j (y_{ij} - \bar{y}_i)^2$	$N(1-\eta^2) S_2^2/(N-h)$
Total	$N-1$	$\sum_i \sum_j (y_{ij} - \bar{y})^2$	—

For testing whether the second mean square is significantly greater than the third, we have the variance ratio

$$F = \frac{\eta^2 - r^2}{1 - \eta^2} \times \frac{N-h}{h-2}; \qquad \nu_1 = h-2, \quad \nu_2 = N-h.$$

If the value of F is significant, the assumption of linearity of regression is discredited. It should be observed that the value of F depends not only on the difference $\eta^2 - r^2$, but also on η^2, N and h. Thus a knowledge of $\eta^2 - r^2$ is by itself insufficient to decide the question of linearity of regression.

Example. A random sample of 200 pairs of values from a bivariate normal population, when grouped in 10 arrays of y's, gave values $r = 0.3$ and $\eta_y = 0.4$. Are these results consistent with the assumption of linearity of regression?

Here $N = 200$, $h = 10$, $\nu_1 = 8$, $\nu_2 = 190$ and

$$F = \frac{0.07}{0.84} \times \frac{190}{8} = 1.98.$$

This value is just on the 5 % level of significance. The assumption of linearity of regression is thus rather discredited.

ANALYSIS OF COVARIANCE

105. Resolution of the 'sum of products'. One criterion of classification

Analysis of covariance has been described as 'the technique of testing for homogeneity in problems dealing with two or more correlated variables'.* Its main use is to test the significance of the difference between the mean values of a variate y in certain classes, when these have been corrected for differences in some concomitant variable x. The method of doing this will be explained shortly; but we must first study the necessary algebraical tools.

The resolution of the total variation into components, already studied in analysis of variance, has its counterpart in analysis of covariance; and the algebra of the two processes runs along parallel lines. The total covariation of a bivariate sample, represented by the sum of the products of the deviations of the variates from their means, may be resolved into components associated with different factors; and from these components, and the corresponding components of variation of x and of y, estimates of the coefficient of regression (or correlation) in the population are determined. The estimate from 'error' is tested for significance as in §§ 103, 89 or 90; and, if this proves to be significant, the other estimates are tested by comparison with it. We are thus able to estimate the effects of the various factors on the degree of association of the variates.

Suppose that the data consist of N pairs of corresponding values of the two variates, x and y, and that these may be grouped in h different classes according to a certain criterion. With the double

* Wishart and Sanders, 1936, 3, p. 46.

suffix notation, (x_{ij}, y_{ij}) is the jth pair of values in the ith class $(i = 1, ..., h)$. The numbers of pairs in the different classes are not necessarily equal. Let n_i be the number of pairs in the ith class, so that $\sum_i n_i = N$. As before let $\bar{x}$, $\bar{y}$ denote the general means of the two variables, and $\bar{x}_i$, $\bar{y}_i$ the means of their values in the ith class. Then

$$\sum_j (x_{ij} - \bar{x}_i) = 0 = \sum_j (y_{ij} - \bar{y}_i). \tag{16}$$

The deviations of x_{ij} and y_{ij} from their general means are expressible as

$$x_{ij} - \bar{x} = (x_{ij} - \bar{x}_i) + (\bar{x}_i - \bar{x}),$$

$$y_{ij} - \bar{y} = (y_{ij} - \bar{y}_i) + (\bar{y}_i - \bar{y}).$$

On forming their product, and summing over all pairs of values, we have

$$\sum_i \sum_j (x_{ij} - \bar{x})(y_{ij} - \bar{y}) = \sum_i \sum_j (x_{ij} - \bar{x}_i)(y_{ij} - \bar{y}_i) + \sum_i n_i(\bar{x}_i - \bar{x})(\bar{y}_i - \bar{y}), \tag{17}$$

the remaining sums disappearing in virtue of (16). Corresponding to this we have also the resolution of the sum of squares of the x's, expressed by (3), and that of the y's given by a similar formula.

Suppose now that the N pairs of values constitute a random sample from a homogeneous population, in which the covariance is μ_{11}. Then, by § 61, the expected value of the sum in the first member of (17) is $(N-1)\mu_{11}$. Similarly, since the values in the ith class may be regarded as a random sample of n_i pairs,

$$E(\sum_j (x_{ij} - \bar{x}_i)(y_{ij} - \bar{y}_i)) = (n_i - 1)\mu_{11},$$

so that

$$E(\sum_i \sum_j (x_{ij} - \bar{x}_i)(y_{ij} - \bar{y}_i)) = \sum_i (n_i - 1)\mu_{11} = (N - h)\mu_{11}.$$

Consequently, since the expectations of the two members of (17) must be equal, that of the final sum in the equation must be $(h-1)\mu_{11}$. Thus the various sums in (17), divided by $N-1$, $N-h$ and $h-1$ respectively, give unbiased estimates of the covariance of the population, based on numbers of D.F. represented by these divisors.

Following Wishart and Sanders* we may conveniently denote the three sums in (17) by C'', C' and C respectively, the equation being then equivalent to $C'' = C' + C$. The corresponding sums of squares for the x's will be denoted by A'', A' and A, and those for the y's by B'', B' and B. Then (3) and its counterpart are equivalent to

$$A'' = A' + A, \quad B'' = B' + B.$$

The resolution of sums of squares and products may be combined in a single table as follows:

Source of variation	D.F.	Sums of squares	Sum of products	Coefficient of regression	Coefficient of correlation
Between classes	$h-1$	A, B	C	$b = C/A$	$r = C/\sqrt{(AB)}$
Within classes	$N-h$	A', B'	C'	$b' = C'/A'$	$r' = C'/\sqrt{(A'B')}$
Total	$N-1$	A'', B''	C''	—	—

We thus obtain different estimates of the coefficients of regression and correlation. By means of §§ 103, 89 or 90 we first test whether the estimate obtained within classes is significant of correlation in the population. If it proves to be significant, we proceed to test the differences between the class means after these have been corrected for regression. Incidentally we may also test the significance of the difference between b and b'.

106. Calculation of the sums of products

The sum of products of the deviations of N pairs of numbers x_s, y_s ($s = 1, ..., N$), from their means $\bar{x}, \bar{y}$ is given by

$$\sum_s (x_s - \bar{x})(y_s - \bar{y}) = \sum_s x_s y_s - \bar{x} \sum_s y_s - \bar{y} \sum_s x_s + N\bar{x}\bar{y}$$
$$= \sum_s x_s y_s - TT'/N, \tag{18}$$

where $\qquad T = \sum x_s = N\bar{x}, \quad T' = \sum y_s = N\bar{y}.$

Applying this formula to the sums in (17) we have

$$C'' = \sum_i \sum_j (x_{ij} - \bar{x})(y_{ij} - \bar{y}) = \sum_i \sum_j x_{ij} y_{ij} - TT'/N, \tag{19}$$

* Wishart and Sanders, 1936. 3. p. 48.

T, T' being the grand totals for x and y respectively. Similarly, on summing first for the values in the ith class, we have

$$C' = \sum_i \sum_j (x_{ij} - \bar{x}_i)(y_{ij} - \bar{y}_i) = \sum_i (\sum_j x_{ij} y_{ij} - T_i T'_i/n_i),$$

where $T_i = \sum_j x_{ij} = n_i \bar{x}_i, \quad T'_i = \sum_j y_{ij} = n_i \bar{y}_i,$

these being the sums of the values of x and y in the ith class. Consequently

$$C' = \sum_i \sum_j x_{ij} y_{ij} - \sum_i T_i T'_i/n_i. \tag{20}$$

Subtraction of (20) from (19) gives the remaining sum

$$C = \sum_i T_i T'_i/n_i - TT'/N, \tag{21}$$

which may, of course, be obtained independently.

Since the deviations from the means are independent of the choice of origin, the results obtained by using (19), (20) and (21) are unaltered by a change of the origins of x and y. In other words, if all the values x_{ij} are decreased (or increased) by the same constant, and all the values y_{ij} by another constant, the results obtained for the three sums of products are unchanged. The arithmetic may often be simplified in this manner, large numbers being replaced by much smaller ones.

107. Examination and elimination of the effect of regression

The method of § 103, for testing the significance of an estimated regression, consists in resolving the total sum of squares of the y's into two components, one due to regression and the other to deviations from the regression line, and then comparing the estimates of population variance derived from these two sums. It will be observed that the test was applied to a sample without any attempt at classification, that is to say, quite apart from any resolution of the variation into components due to factors other than regression. In terms of the total sums of squares and products, A'', B'', C'', the sum of squares due to regression is expressible as

$$NS_2^2 r^2 = B''(C''^2/A''B'') = C''^2/A'', \tag{22}$$

and that due to deviation from the regression line is $B'' - C''^2/A''$. The former corresponds to 1 D.F. and the latter to $N-2$, which is one less than for the total sum of squares.

Now we have seen how, after the values in the sample have been classified, we may resolve the total variation of the variables, and their covariation, into components between the means of classes and within classes respectively. From each set of components we may calculate an estimate of the regression coefficient and a line of regression; and the corresponding variation of the y's may be further resolved by the method of § 103 into a part due to regression, and another due to deviation from the regression line, the number of D.F. for the latter being one less than the total D.F. for that component. Applying this process to each line of the table at the end of § 105, we find from each a sum of squares of deviations from the corresponding line of regression, with D.F. as indicated in the following table:*

Source of variation	D.F.	Residual sum of squares
Between classes	$h-2$	$B-C^2/A$
Within classes	$N-h-1$	$B'-C'^2/A'$
Total	$N-2$	$B''-C''^2/A''$

leading to different estimates of the variance of y in the population. Denoting the estimate within classes by s^2 we have

$$s^2 = \frac{B' - C'^2/A'}{N-h-1}. \tag{23}$$

This is the estimate of the variance of y after correction for regression. The significance of any other estimate of the variance is tested by comparison with it.

The significance of any apparent regression is first tested by the method of § 103, applied to the sums within classes; that is to say, we compare the estimate C'^2/A' of variance, due to regression and based on 1 D.F., with the above estimate s^2 based on $N-h-1$ D.F. If the regression proves to be significant we proceed to test the differences between class means after correction for regression. Subtracting the second row of the above table from the third we have the sum of squares $B + C'^2/A' - C''^2/A''$ with $h-1$ D.F., which we also compare with the sum of squares within classes. If it proves

* Cf. Wishart and Sanders, 1936, 3, p. 49.

to be significant we conclude that there are differences between the class means, after these have been adjusted by the regression coefficient b' within classes.

Now in the first line of the above table we have a sum of squares between class means, adjusted by the regression coefficient b, and corresponding to $(h-2)$ D.F. The testing of this sum in place of the above raises the question of the significance of the difference between b and b'. This may be examined as follows. Tabulating the two sums of squares just mentioned, and their D.F., we may arrange the work:

Between classes	D.F.	Sum of squares
Adjusted	$h-1$	$B + C'^2/A' - C''^2/A''$
Adjusted by b	$h-2$	$B - C^2/A$
Difference	1	$C^2/A + C'^2/A' - C''^2/A''$

Now the sum in the last row may be expressed

$$\frac{C^2}{A} + \frac{C'^2}{A'} - \frac{(C+C')^2}{A+A'} = \frac{AA'}{A+A'}\left(\frac{C}{A} - \frac{C'}{A'}\right)^2 = \frac{AA'(b-b')^2}{A+A'}.$$

Comparing this estimate of variance, based on 1 D.F., with the estimate s^2 based on $N-h-1$, we have the variance ratio

$$F = \frac{AA'(b-b')^2}{A+A'} \times \frac{N-h-1}{B'-C'^2/A'}, \tag{24}$$

$$\nu_1 = 1, \quad \nu_2 = N-h-1.$$

A rare value of F indicates that the two coefficients, b and b', are significantly different; in which case the factor of classification has an effect upon the degree of association of the variables.

Example. To examine the relation between the yield of grain (x bushels/acre) and the cost of production (y shillings/bushel) in a certain state, six districts were chosen at random, and in each a random selection of five farms was made. The results for the season are tabulated below, columns corresponding to districts. Is there any significant indication of correlation between the variables; and, if so, does its value vary with the district?

x	y	x	y	x	y	x	y	x	y	x	y
13	3·5	10	4·7	16	2·8	9	4·5	12	3·8	15	4·7
11	4·0	9	5·3	13	3·0	12	3·6	17	2·2	11	5·1
18	2·5	8	5·6	15	2·8	10	4·0	19	2·0	9	6·0
14	3·7	12	4·0	10	4·2	13	3·4	11	4·6	14	4·3
12	4·3	11	4·4	12	3·2	8	5·5	14	3·4	17	3·9

Shifting the origin to $x = 12$, $y = 3$, and rewriting the table, the student will easily verify that

$$T_i = 8, \quad -10, \quad 6, \quad -8, \quad 13, \quad 6; \qquad T = 15,$$
$$T'_i = 3, \quad\quad 9, \quad 1, \quad\quad 6, \quad 1, \quad 9; \qquad T' = 29,$$

and hence that

$$T^2/N = 7{\cdot}5, \quad T'^2/N = 28{\cdot}03, \quad TT'/N = 14{\cdot}5, \quad \sum_i T_i T'_i/5 = -8{\cdot}2,$$

$$\sum_i \sum_j x_{ij}^2 = 259, \quad \sum_i \sum_j y_{ij}^2 = 56{\cdot}96, \quad \sum_i \sum_j x_{ij}y_{ij} = -54{\cdot}8.$$

Consequently

$$A = 93{\cdot}8 - 7{\cdot}5 = 86{\cdot}3, \quad A' = 259 - 93{\cdot}8 = 165{\cdot}2,$$
$$A'' = 259 - 7{\cdot}5 = 251{\cdot}5.$$

$$B = 41{\cdot}8 - 28{\cdot}03 = 13{\cdot}77, \quad B' = 56{\cdot}96 - 41{\cdot}8 = 15{\cdot}16,$$
$$B'' = 56{\cdot}9 - 28{\cdot}03 = 28{\cdot}93.$$

$$C = -8{\cdot}2 - 14{\cdot}5 = -22{\cdot}7, \quad C' = -54{\cdot}8 + 8{\cdot}2 = -46{\cdot}6,$$
$$C'' = -54{\cdot}8 - 14{\cdot}5 = -69{\cdot}3.$$

The tabulated results are:

Source of variation	D.F.	Sum of squares		Sums of products	Coefficient of regression
		(x^2)	(y^2)		
Between districts	5	86·3	13·77	$-22{\cdot}7$	$-0{\cdot}263$
Within districts	24	165·2	15·16	$-46{\cdot}6$	$-0{\cdot}282$
Total	29	251·5	28·93	$-69{\cdot}3$	$-0{\cdot}276$

First, to test the significance of b' we have, as explained above,

$$F = \frac{C'^2/A'}{B' - C'^2/A'} \times (N - h - 1) = \frac{13{\cdot}14 \times 23}{2{\cdot}02} = 149^{**}$$

$$\nu_1 = 1, \quad \nu_2 = 23.$$

Thus the value of b' is highly significant of negative correlation.

To test the differences of class means, after these have been corrected for regression, we calculate the estimate of variance

$$\frac{B + C'^2/A' - C''^2/A''}{h - 1} = \frac{13{\cdot}77 + 13{\cdot}14 - 19{\cdot}09}{5} = 1{\cdot}56,$$

and compare this with the estimate within classes

$$s^2 = 2{\cdot}02/23 = 0{\cdot}088.$$

The variance ratio is

$$F = 1\cdot56/0\cdot088 = 17\cdot7**$$

which, for $\nu_1 = 5$ and $\nu_2 = 23$, is highly significant. We conclude that the cost of production, corrected for differences in yield, varies significantly from district to district.

By using (24) show that b and b' do not differ significantly.

108. Two criteria of classification

Suppose next that the N pairs of values (x_{ij}, y_{ij}) in the sample may be classified according to two different criteria, H and K, the former determining h different classes and the latter k different groups. Suppose for simplicity that one pair of values from each group is present in each class, and vice versa, so that $N = hk$. The pairs of values may be arranged in a rectangular array of h columns and k rows, in which columns correspond to classification H and rows to K. The pair (x_{ij}, y_{ij}) belongs to the ith class and the jth group, and appears in the ith column and the jth row.

The resolution of the sum of products expressed in (17) is still valid. But the first sum in the second member of the equation may be further resolved. Thus if

$$X_{ij} = x_{ij} - \bar{x}_i, \quad Y_{ij} = y_{ij} - \bar{y}_i,$$

and we apply the resolution expressed by (17) to the sum of products $\sum_i \sum_j X_{ij} Y_{ij}$, making it however with respect to the means of rows instead of the means of columns, we find by the same argument as in § 100 that

$$\sum_i \sum_j (x_{ij} - \bar{x})(y_{ij} - \bar{y}) = \sum_i k(\bar{x}_i - \bar{x})(\bar{y}_i - \bar{y}) + \sum_j h(\bar{x}_j - \bar{x})(\bar{y}_j - \bar{y})$$
$$+ \sum_i \sum_j (x_{ij} - \bar{x}_i - \bar{x}_j + \bar{x})(y_{ij} - \bar{y}_i - \bar{y}_j + \bar{y}). \quad (25)$$

Denoting the various sums in this equation by C'', C_1, C_2, C' we may write it briefly

$$C'' = C_1 + C_2 + C'.$$

On the assumption that the pairs of values constitute a random sample from a homogeneous population, the expected value of the first member of (25) is $(N-1)\mu_{11}$, where μ_{11} is the covariance in the population. That of the first sum in the second member has been shown to be $(h-1)\mu_{11}$, and similarly that of the second sum is $(k-1)\mu_{11}$. Consequently the expectation of the final sum is

$$(N-h-k+1)\mu_{11} = (h-1)(k-1)\mu_{11}.$$

On dividing the various sums by $(N-1)$, $(h-1)$, $(k-1)$ and $(h-1)(k-1)$ respectively, we thus obtain four unbiased estimates of the covariance of the homogeneous population, with D.F. represented by these divisors. In §100 we obtained the corresponding estimates of the variance of x, and similar equations give estimates of the variance of y in the population. From the partial sums we obtain three independent estimates of the coefficients of regression and correlation. Denoting the sums of squares of the x's by A'', A_1, A_2, A' and those of the y's by B'', B_1, B_2, B', we may tabulate the results:

Source of variation	D.F.	Sums of squares		Sum of products	Coefficient of regression
		(x^2)	(y^2)		
Between classes	$h-1$	A_1	B_1	C_1	$b_1 = C_1/A_1$
Between groups	$k-1$	A_2	B_2	C_2	$b_2 = C_2/A_2$
Error	$(h-1)(k-1)$	A'	B'	C'	$b' = C'/A'$
Total	$hk-1$	A''	B''	C''	—

The first step is to test the significance of the regression coefficient b' obtained from the error sums; and this is done exactly as in §107, except that the number of D.F. for error is here $(h-1)(k-1)$. Thus the residual sum of squares due to error is $B' - C'^2/A'$, with $N-h-k$ D.F., giving the estimate of variance of y

$$s^2 = (B' - C'^2/A')/(N-h-k).$$

With this we compare the estimate C'^2/A' obtained from the regression sum of squares corresponding to 1 D.F., and the result settles whether b' is significant. If it is, we proceed to test the significance of the differences of the means of y in classes (or groups) after these have been corrected for regression on x. This may be done as follows. Take the first and third rows of the above table, and by addition form a combined source of variation, $S = (\text{classes} + \text{error})$, with sums of squares and products A, B, C. Thus we have:

Source of variation	D.F.	Sums of squares and products
Classes	$h-1$	$A_1,\ B_1,\ C_1$
Error	$N-h-k+1$	$A',\ B',\ C'$
Total, S	$N-k$	$A,\ B,\ C$

where $A = A_1 + A'$, etc. From these we form the corresponding residual sums of squares as in § 107, viz.

Source of variation	D.F.	Residual sum of squares
Classes	$h-2$	$B_1 - C_1^2/A_1$
Error	$N-h-k$	$B' - C'^2/A'$
S	$N-k-1$	$B - C^2/A$

Subtracting the second row from the third we obtain a residual sum of squares $B_1 + C'^2/A' - C^2/A$ corresponding to $h-1$ D.F. Comparing the estimate of variance obtained from this with the estimate s^2 we are able to decide whether the class means differ significantly after correction for regression.

Incidentally, we may deduce a test for the significance of the difference $b_1 - b'$. For, on subtracting the residual sums of squares for classes and error from that for S, we have the sum

$$C_1^2/A_1 + C'^2/A' - C^2/A$$

with 1 D.F. As in § 107 this sum is expressible as

$$A_1 A'(b_1 - b')^2/(A_1 + A');$$

so that in testing it we are testing the difference $b_1 - b'$. Comparing it with the residual sum of squares for error we have a variance ratio

$$F = \frac{A_1 A'(b_1 - b')^2}{A_1 + A'} \times \frac{N-h-k}{B' - C'^2/A'}, \qquad (26)$$

$$\nu_1 = 1, \quad \nu_2 = N-h-k.$$

We thus determine whether b_1 is significantly different from b'; and we may test b_2 by a similar formula.*

* For a numerical illustration see Ex. XI, 15, below.

COLLATERAL READING

FISHER, 1938, 2, chapters VII and VIII.
WISHART, 1934, 3; 1936, 2; 1940, 5.
WISHART and SANDERS, 1936, 3.
TIPPETT, 1931, 2, chapters VI, VII, IX, X, XI.
RIDER, 1939, 6, chapters VIII and IX.
GOULDEN, 1939, 2, chapters XI, XII, XIII and XV.
MILLS, 1938, 1, chapter XV and pp. 681–90.
SNEDECOR, 1934, 2; 1938, 3, chapters X and XI.
YATES, 1934, 5; 1937, 4 and 1938, 4.
IRWIN, 1931, 1 and 1934, 4.

EXAMPLES XI

1. Verify as follows that the expected value of the final sum in § 97 (3) is $(h-1)\sigma^2$. If μ is the mean of the population,

$$\sum_i n_i(\bar{x}_i - \bar{x})^2 = \sum_i n_i(\bar{x}_i - \mu)^2 - N(\bar{x} - \mu)^2.$$

Since $E(\bar{x}_i - \mu)^2$ is the variance of the mean of a sample of n_i members, and $E(\bar{x} - \mu)^2$ is that of the mean of a sample of N members, it follows that

$$E\left[\sum_i n_i(\bar{x}_i - \bar{x})^2\right] = \sum_i n_i(\sigma^2/n_i) - N\sigma^2/N = (h-1)\sigma^2.$$

Verify the mean value of $\sum_i \sum_j (x_{ij} - \bar{x}_i)^2$ similarly.

2. With the notation of § 100, $Y_{ij} = x_{ij} - \bar{x}_i - \bar{x}_j + \bar{x}$, show that $\sum_i Y_{ij} = 0 = \sum_j Y_{ij}$. Thus the sum of the Y's in any row or in any column is zero. Only $(h-1)(k-1)$ of the quantities Y_{ij} are independent.

3. To test the significance of the variation of the retail price of a certain commodity among four large cities, A, B, C and D, seven shops were chosen at random in each city, and the prices observed were as follows:

A: 6s. 10d., 6s. 7d., 6s. 1d., 5s. 9d., 5s. 9d., 5s. 3d., 5s. 1d.

B: 7s., 6s. 10d., 6s. 8d., 6s. 7d., 6s. 4d., 5s. 8d., 5s. 2d.

C: 7s. 4d., 7s., 6s. 8d., 5s. 8d., 5s. 8d., 5s. 6d., 5s. 6d.

D: 6s. 7d., 6s. 5d., 6s. 4d., 6s. 2d., 6s., 5s. 8d., 5s. 4d.

Do the data indicate that the prices for the four cities are significantly different?

The classes correspond to the cities. Expressing the prices in pence show that the tabulated results are:

Source of variation	D.F.	Sum of squares	Mean square
Between cities	3	94·96	31·65
Within cities	24	1446·00	60·25
Total	27	1540·96	—

Thus the mean square between cities is less than that within cities, and is therefore not significantly large. Neither is it significantly small because, for $\nu_1 = 24$ and $\nu_2 = 3$, a value of F in the neighbourhood of 2 is not significant.

Show that the s.e. of the difference of the means for two cities is $4.15d.$; and hence for no two cities is the difference of the mean prices significant.

4. Show that, if $\bar{x}_1$ and $\bar{x}_2$ are the means of two classes of n_1 and n_2 members respectively, and s^2 is the estimate of population variance derived from error and corresponding to ν D.F., then $s\sqrt{(1/n_1 + 1/n_2)}$ is the s.e. of the difference of the means of the two classes, and hence that the statistic

$$t = \frac{(\bar{x}_1 - \bar{x}_2)}{s\sqrt{(1/n_1 + 1/n_2)}}$$

is distributed like t for ν D.F. This formula provides a method of testing the significance of the difference of the means of two unequal classes.

5. In a certain large country, to test the variation in the price of a certain commodity with district and season, six districts were chosen at random, and the price was observed in each on six random occasions throughout the year. The observations are recorded in pence in the table below, in which columns correspond to seasons and rows to districts. Discuss the significance of the variation with season and with district.

$$
\begin{array}{cccccc}
11 & 15 & 24 & 19 & 21 & 17 \\
12 & 16 & 25 & 18 & 22 & 16 \\
10 & 13 & 21 & 17 & 20 & 15 \\
9 & 14 & 18 & 18 & 21 & 13 \\
13 & 16 & 20 & 16 & 23 & 14 \\
14 & 17 & 23 & 20 & 22 & 15 \\
\end{array}
$$

Show that analysis of variance leads to the following results:

Source of variation	D.F.	Sum of squares	Mean square	F
Seasons	5	492·33	98·46	55·5**
Districts	5	44·33	8·86	5·0**
Error	25	44·33	1·77	—
Total	35	581·00	—	—

Both the variance ratios are highly significant. Show that the s.e. of the difference of the means for two seasons or two districts is 0·77; and deduce that the prices in the first, second and last districts do not differ significantly.

6. An agricultural experiment, on the Latin square plan, gave the following results for the yield of wheat per acre, letters corresponding to varieties, columns to treatments and rows to blocks. Discuss the variation of yield with each of the factors:

A 16	*B* 10	*C* 11	*D* 9	*E* 9
E 10	*C* 9	*A* 14	*B* 12	*D* 11
B 15	*D* 8	*E* 8	*C* 10	*A* 18
D 12	*E* 6	*B* 13	*A* 13	*C* 12
C 13	*A* 11	*D* 10	*E* 7	*B* 14

Show that the results obtained by analysis of variance are:

Source of variation	D.F.	Sum of squares	Mean square	F
Treatments	4	66·56	16·64	37·8**
Blocks	4	2·16	0·54	1·2
Varieties	4	122·56	30·64	69·6**
Error	12	5·28	0·44	—
Total	24	196·56	—	—

Show also that the s.e. of the difference of the means of two classes is 0·42; and hence that for no two blocks do the means differ significantly.

7. A random sample of 100 pairs of values from a normal population, grouped in 10 arrays of y's, gave a correlation ratio of y on x equal to 0·3. Show that this value is not significant of association between the variables.

8. A random sample of 150 pairs from a bivariate normal population when grouped in 15 arrays of y's gave values $r = 0·4$ and $\eta_y = 0·5$. Show that these results are consistent with the assumption of linearity of regression of y on x.

9. A parabola, fitted to a random sample of 45 pairs of values from a normal population, gave an index of correlation $R = 0·3$. Show that this value is not significant of parabolic regression. Also show that it would not be significant for a sample of less than 83 pairs.

10. Give a direct proof that the expected value of the last sum in § 105 (17) is $(h-1)\mu_{11}$.

11. Give a direct proof of the formula § 106 (21) for C.

12. In the example of § 107, show that the sum of squares represented by $B - C^2/A$, which corresponds to 4 D.F., is highly significant. Hence the deviations of the means of districts from the line of regression fitted to them are too great to be attributed to chance.

13. Supply the details of the proof of § 108 (25).

14. With the notation of § 107 the difference of the means of the y's for the pth and qth classes, when corrected for regression, is

$$\bar{y}_p - \bar{y}_q - b'(\bar{x}_p - \bar{x}_q) \tag{i}$$

which consists of two independent parts. The estimated variance of the first part is $2s^2/k$, where k is the number n_i in each class, and s^2 is the error mean square $(B' - C'^2/A')/(N - h - 1)$ in the analysis of residual variance. The estimated variance of b' is s^2/A'. Hence show that the estimated variance of the difference (i) is

$$s^2[2/k + (\bar{x}_p - \bar{x}_q)^2/A']. \tag{ii}$$

The quotient of the difference (i) by the square root of the expression (ii) is distributed like t with $N - h - 1$ D.F. (Cf. Wishart, 1936, 2; also Wishart and Sanders, 1936, 3, pp. 53–4.)

15. Examine, as in § 108, the covariation between the yields of grain and straw, as indicated by the following data.* Each of 5 blocks was divided into 5 plots, and 5 different treatments (A, B, ..., E) were distributed at random among the plots of each block. The columns correspond to blocks, and the yields of grain and straw are denoted by x and y respectively:

	x	y	x	y	x	y	x	y	x	y
A:	65	32	68	26	65	32	71	26	62	33
B:	75	38	54	20	71	32	71	28	64	29
C:	72	33	69	30	69	38	69	30	61	30
D:	70	29	69	27	72	26	70	24	70	31
E:	70	37	67	29	52	33	66	23	66	40

The factors of classification are blocks and treatments. Diminishing all the x's by 66 and all the y's by 26, verify that

$$T_i = 22,\quad -3,\quad -1,\quad 17,\quad -7; \qquad T = 28$$
$$T'_i = 39,\quad 2,\quad 31,\quad 1,\quad 33; \qquad T' = 106,$$

and calculate the values of T_j and T'_j. Hence show that the sums of squares and products are given by:

Source of variation	D.F.	Sum of squares		Sum of products	Coefficient of regression
		(x^2)	(y^2)		
Blocks	4	135·04	265·76	2·68	0·020
Treatments	4	98·24	87·36	−64·12	−0·653
Error	16	455·36	211·44	174·72	0·384
Total	24	688·64	564·56	113·28	—

The estimate s^2 obtained from the residual sum of squares for error is

$$s^2 = \frac{B' - C'^2/A'}{N - h - k} = \frac{144·34}{15} = 9·62.$$

To test the significance of b' we compare with s^2 the estimate $C'^2/A' = 67·1$ corresponding to 1 D.F. The variance ratio is

$$F = 67·1/9·62 = 6·9 \quad (\nu_1 = 1,\ \nu_2 = 15).$$

* Data adapted from Fisher and Eden, *Journ. Agric. Sci.*, vol. 17 (1927), p. 548.

Thus the value of b' is decidedly significant of positive correlation between x and y, apart from the factors of classification. To test whether the yield of straw, corrected for yield of grain, varies significantly with the treatment, we calculate

$$C = C_2 + C' = 110{\cdot}6, \quad A = A_2 + A' = 553{\cdot}6$$

and compare the estimate of variance $(B_2 + C'^2/A' - C^2/A)/(k-1)$ based on 4 D.F. with s^2 based on 15. The former has the value

$$(87{\cdot}36 + 67{\cdot}1 - 22{\cdot}1)/4 = 33{\cdot}09,$$

and the latter is 9·62. The variance ratio is

$$F = 33{\cdot}09/9{\cdot}62 = 3{\cdot}44$$

which, for $\nu_1 = 4$ and $\nu_2 = 15$, is significant at the 5 % level. We conclude that the yield of straw, after correction for yield of grain, does vary significantly with the treatment.

Show that the difference $b' - b_2$, tested by § 108 (26), is decidedly significant.

CHAPTER XII

MULTIVARIATE DISTRIBUTIONS.
PARTIAL AND MULTIPLE CORRELATIONS

109. Introductory. Yule's notation

In considering a bivariate distribution we saw that, when it is known that the values of one variable are influenced by those of another, the coefficient of correlation provides a useful measure of the degree of association between them. But it often happens that the values of a variable are influenced by those of several others. It is known, for instance, that the statures of men are influenced by those of their ancestors; and the yield of grain is affected by the amounts of different fertilizers used. In such cases our data usually constitute a distribution of values of several variables. If we are concerned with the combined influence of a group of variables upon a variable not included in that group, our study is that of multiple regression and *multiple correlation*. If, however, we wish to examine the effect of one variable upon a second, after eliminating the effects of other variables, our problem is that of *partial correlation*. The analysis involved is rendered simple and compact by a notation due to Yule,* which has gained a fairly wide acceptance in recent years. Before applying this to the more general case of several variables, we shall pave the way for the student by reviewing the results obtained for two variables in Chapter IV, and expressing them in Yule's notation.

Let the two variables, *measured from their means*, be x_1 and x_2, with standard deviations σ_1 and σ_2. If the lines of regression of x_1 on x_2, and of x_2 on x_1, are

$$x_1 = b_{12}x_2, \quad x_2 = b_{21}x_1, \tag{1}$$

respectively, the *residuals*, or errors of estimate of the variables incurred by using these equations, are expressed by

$$x_{1.2} = x_1 - b_{12}x_2, \quad x_{2.1} = x_2 - b_{21}x_1. \tag{2}$$

* Yule, 1907, 1.

These residuals are the *deviations* of the representative points from the corresponding line of regression. The values of the coefficients of regression, b_{12} and b_{21}, are obtained by minimizing the sums of squares of the residuals; and this is done by equating to zero the partial derivatives of these sums with respect to b_{12} and b_{21}, and the constant terms. This leads to the *normal equations*

$$\sum(x_1 - b_{12}x_2) = 0, \quad \sum x_2(x_1 - b_{12}x_2) = 0 \tag{3}$$

in the one case, and

$$\sum(x_2 - b_{21}x_1) = 0, \quad \sum x_1(x_2 - b_{21}x_1) = 0 \tag{3'}$$

in the other. The first equation in each case expresses the fact that the mean of the distribution is on the line of regression; and, with the mean as origin, the equations of these lines have the simple form (1). The above normal equations, expressed in the more compact notation, are

$$\sum x_{1.2} = 0, \quad \sum x_2 x_{1.2} = 0 \tag{3}$$

and
$$\sum x_{2.1} = 0, \quad \sum x_1 x_{2.1} = 0, \tag{3'}$$

respectively, the summation including all pairs of values of the distribution. It will be observed that, whereas in Chapter IV subscripts were used to distinguish the pairs of values in the distribution, they are here used to distinguish the different variables. They are no longer needed for the former purpose, since $\sum$ will throughout denote summation over the whole distribution.

From (3) and (3') we have the familiar values of the coefficients of regression,

$$b_{12} = \sum x_1 x_2 / \sum x_2^2, \quad b_{21} = \sum x_1 x_2 / \sum x_1^2.$$

The coefficient of correlation, which we denote by r_{12} or r_{21}, is such that

$$r_{12}^2 = b_{12} b_{21}, \tag{4}$$

and
$$b_{12} = r_{12}\sigma_1/\sigma_2, \quad b_{21} = r_{21}\sigma_2/\sigma_1, \tag{5}$$

r_{12} having the same sign as b_{12} or b_{21}, which is the sign of $\sum x_1 x_2$. It should be remembered that, although $r_{12} = r_{21}$, the values of b_{12} and b_{21} are in general different. Lastly the mean squares of the deviations (2) are denoted by $\sigma_{1.2}^2$ and $\sigma_{2.1}^2$ respectively. In virtue of §26 (21) and (23), these are given by

$$\left.\begin{aligned}\sigma_{1.2}^2 &= \frac{1}{N}\sum x_{1.2}^2 = \sigma_1^2(1-r_{12}^2),\\[2mm]\sigma_{2.1}^2 &= \frac{1}{N}\sum x_{2.1}^2 = \sigma_2^2(1-r_{21}^2).\end{aligned}\right\} \tag{6}$$

The quantities $\sigma_{1.2}$ and $\sigma_{2.1}$ are referred to as standard deviations of the first order, the order being the number of subscripts following the point. These are called *secondary subscripts*, while *primary subscripts* are those preceding the point. The standard deviations σ_1 and σ_2 are of zero order. The residuals $x_{1.2}$ and $x_{2.1}$ are deviations of the first order.

The regression of x_1 on x_2 is said to be *linear*, if the mean of each array of x_1's is on the line of regression.

110. Distribution of three or more variables

Consider next a distribution of three variables which, measured from their means, are x_1, x_2 and x_3. The data consist of N sets of corresponding values of the three variables. We enquire first as to the best estimate of x_1 that can be obtained from the data in the form of a linear function of x_2 and x_3. The regression equation is thus of the form

$$x_1 = a + b_{12.3}x_2 + b_{13.2}x_3.$$

The 'best' estimate is interpreted in accordance with the principle of least squares, the constants being chosen so as to make the sum of the squares of the errors of estimate a minimum. The subscripts to the coefficients are written down on the following principle. The first is that of the variable for which the estimate is being found, and the second is that of the variable which the coefficient multiplies. These are primary subscripts. Separated from them by a point are the subscripts of the other variables that enter into the equation. These are secondary subscripts, and their number determines the *order* of the regression coefficient. In the above equation the coefficients are partial regression coefficients of the first order. The significance of the term 'partial' will be explained shortly. The constant a has been given no subscripts, because it will appear immediately that its value is zero.

The sum of the squares of the residuals to be minimized is

$$\sum (x_1 - a - b_{12.3}x_2 - b_{13.2}x_3)^2.$$

Equating to zero its partial derivatives with respect to a and the b's we have the *normal equations*

$$\sum (x_1 - a - b_{12.3}x_2 - b_{13.2}x_3) = 0,$$
$$\sum x_2(x_1 - a - b_{12.3}x_2 - b_{13.2}x_3) = 0,$$
$$\sum x_3(x_1 - a - b_{12.3}x_2 - b_{13.2}x_3) = 0.$$

Since the mean of each variable is zero, the first of these equations gives $a = 0$; and the regression equation of x_1 on x_2 and x_3 is simply

$$x_1 = b_{12.3}x_2 + b_{13.2}x_3. \tag{7}$$

The other two normal equations may then be written more concisely

$$\sum x_2 x_{1.23} = 0 = \sum x_3 x_{1.23}, \tag{8}$$

in which $x_{1.23}$, defined by

$$x_{1.23} = x_1 - b_{12.3}x_2 - b_{13.2}x_3, \tag{9}$$

is the residual, or error of estimate of x_1 from the regression equation (7). Its mean is zero, since those of the other variables are zero. Hence the s.d. of this residual, denoted by $\sigma_{1.23}$, is given by

$$N\sigma_{1.23}^2 = \sum x_{1.23}^2, \tag{10}$$

the summation covering the whole distribution, whose total frequency is N. This is a s.d. of order two, since it has two secondary subscripts.

Similarly, we have the regression equation of x_2 on x_1 and x_3, namely,

$$x_2 = b_{21.3}x_1 + b_{23.1}x_3, \tag{11}$$

for which the error of estimate is

$$x_{2.13} = x_2 - b_{21.3}x_1 - b_{23.1}x_3, \tag{12}$$

and the normal equations

$$\sum x_1 x_{2.13} = 0 = \sum x_3 x_{2.13}. \tag{13}$$

There are also similar equations for the regression of x_3 on x_1 and x_2. In general the coefficients $b_{12.3}$ and $b_{21.3}$ are different.

The normal equations (3), (8) and (13) express that *the sum of the products of corresponding values of a variable and a residual is zero, when the subscript of the variable is included among the secondary subscripts of the residual,* the summation covering the whole distribution. Further, if $x_{1.2}$ is the residual defined by (2), we have

$$\sum x_{1.23} x_{1.2} = \sum x_{1.23}(x_1 - b_{12} x_2) = \sum x_{1.23} x_1.$$

Similarly,

$$\sum x_{1.23} x_{1.23} = \sum x_{1.23}(x_1 - b_{12.3} x_2 - b_{13.2} x_3) = \sum x_{1.23} x_1$$

in virtue of (8). Thus it is evident that *the sum of the products of two residuals is unaltered by omitting from one of the factors any secondary subscripts which are common to both.* In virtue of this result and the normal equations it follows that *the sum of the products of two residuals is zero if all the subscripts of the one are included among the secondary subscripts of the other.*

The argument of this section is also applicable to the case of n variables $x_1, x_2, \ldots, x_n$. In the regression equation of x_1 on $x_2, \ldots, x_n$, which corresponds to (7), the coefficients have each $n-2$ secondary subscripts, and are therefore of that order. The residual $x_{1.23\ldots n}$ corresponding to (9) is of order $n-1$, as is also its s.d. $\sigma_{1.23\ldots n}$. The normal equations are

$$\sum x_i x_{1.23\ldots n} = 0 \quad (i = 2, 3, \ldots, n).$$

And the theorem concerning the omission of common secondary subscripts is equally valid in the case of n variables.

111. Determination of the coefficients of regression

The regression equation (7) may be interpreted as representing a plane, called the *plane of regression* of x_1 on x_2 and x_3. The normal equations (8) may be expressed

$$\left.\begin{array}{l} \sigma_1 \sigma_2 r_{12} - b_{12.3} \sigma_2^2 - b_{13.2} \sigma_3 \sigma_2 r_{32} = 0, \\ \sigma_1 \sigma_3 r_{13} - b_{12.3} \sigma_2 \sigma_3 r_{23} - b_{13.2} \sigma_3^2 = 0, \end{array}\right\} \quad (14)$$

where r_{12} is the correlation between x_1 and x_2, obtained by ignoring the values of x_3. This is called the *total correlation* of x_1 and x_2. Similarly, r_{23} is the total correlation of x_2 and x_3, and so on. These coefficients are symmetrical in the subscripts. We may eliminate the b's between (7) and (14) by equating to zero the determinant of their coefficients and the remaining terms. Dividing the second and third rows of this determinant by σ_2 and σ_3 respectively, and then the first, second and third columns by σ_1, σ_2 and σ_3 respectively, we obtain the regression equation of x_1 on the other variables in the form

$$\begin{vmatrix} \dfrac{x_1}{\sigma_1} & \dfrac{x_2}{\sigma_2} & \dfrac{x_3}{\sigma_3} \\[2mm] r_{12} & 1 & r_{32} \\[2mm] r_{13} & r_{23} & 1 \end{vmatrix} = 0. \tag{15}$$

If then ω is used to denote the determinant

$$\omega = \begin{vmatrix} 1 & r_{21} & r_{31} \\ r_{12} & 1 & r_{32} \\ r_{13} & r_{23} & 1 \end{vmatrix} \tag{16}$$

and ω_{ij} is the cofactor of the element in the ith column and the jth row, we may write (15) as

$$\omega_{11}\frac{x_1}{\sigma_1}+\omega_{12}\frac{x_2}{\sigma_2}+\omega_{13}\frac{x_3}{\sigma_3} = 0. \tag{17}$$

From this form of the regression equation it follows that

$$b_{12.3} = -\frac{\sigma_1}{\sigma_2}\frac{\omega_{12}}{\omega_{11}}, \quad b_{13.2} = -\frac{\sigma_1}{\sigma_3}\frac{\omega_{13}}{\omega_{11}}. \tag{18}$$

The regression of x_1 on x_2 and x_3 is said to be linear if the mean of each array of x_1's lies on the plane of regression (17).

The residual $x_{1.23}$ may be looked upon as the deviation of the representative point from the plane of regression. The term *deviation* is therefore often used in place of *residual*. The variance $\sigma_{1.23}^2$ of $x_{1.23}$ may be expressed in terms of ω, ω_{11} and σ_1. For

$$N\sigma_{1.23}^2 = \sum x_{1.23}^2 = \sum x_1(x_1 - b_{12.3}x_2 - b_{13.2}x_3).$$

This is equivalent to

$$(\sigma_1^2 - \sigma_{1.23}^2) - b_{12.3}\sigma_1\sigma_2 r_{21} - b_{13.2}\sigma_1\sigma_3 r_{31} = 0,$$

and, on eliminating the b's between this equation and (14), we have a result which may be expressed

$$\begin{vmatrix} 1 - \dfrac{\sigma_{1.23}^2}{\sigma_1^2} & r_{21} & r_{31} \\[2mm] r_{12} & 1 & r_{32} \\[2mm] r_{13} & r_{23} & 1 \end{vmatrix} = 0,$$

or
$$\omega - \omega_{11}\sigma_{1.23}^2/\sigma_1^2 = 0.$$

Consequently
$$\sigma_{1.23}^2 = \omega\sigma_1^2/\omega_{11}. \tag{19}$$

The above argument clearly holds for the more general case of n variables. In place of (16) we then have a determinant ω of order n; and the regression equation of x_1 on $x_2, \ldots, x_n$ is

$$\omega_{11}\frac{x_1}{\sigma_1} + \omega_{12}\frac{x_2}{\sigma_2} + \ldots + \omega_{1n}\frac{x_n}{\sigma_n} = 0. \tag{20}$$

From this we obtain, as in (18), the regression coefficients of order $n-2$. In place of (19) we have an equation giving the variance of the deviation $x_{1.23\ldots n}$, namely,

$$\sigma_{1.23\ldots n}^2 = \omega\sigma_1^2/\omega_{11}. \tag{21}$$

Example. Let x_1, x_2, x_3 in. be the excesses of the heights of father, mother and son respectively above their mean values. A distribution of these variables gave the following approximate correlations and standard deviations:*

$$r_{12} = 0.28, \quad r_{23} = 0.49, \quad r_{31} = 0.51,$$
$$\sigma_1 = 2.7, \quad \sigma_2 = 2.4, \quad \sigma_3 = 2.7.$$

Show that the regression equation of x_3 on x_1 and x_2 is

$$x_3 = 0.40x_1 + 0.42x_2,$$

and deduce that, if the mean heights of father, mother and son are 67·68, 62·48 and 68·65 in. respectively, the regression equation for actual heights X_1, X_2, X_3 is
$$X_3 = 15.3 + 0.40X_1 + 0.42X_2.$$

Also show that $\sigma_{3.12} = 2.1$.

* Modified data from Pearson and Lee, *Biometrika*, vol. 2 (1903), pp. 357–462.

112. Multiple correlation

We proceed to find the correlation between the variable x_1 and its estimate from the regression equation (7). This correlation, which is an indication of the agreement between x_1 and its estimate, is called the *coefficient of multiple correlation* between x_1 and the two variables x_2 and x_3, and is denoted* by $R_{1(23)}$. It may be determined as follows. If $e_{1.23}$ is the estimate of x_1 given by (7), we have

$$e_{1.23} = b_{12.3}x_2 + b_{13.2}x_3$$

or
$$e_{1.23} = x_1 - x_{1.23}. \tag{22}$$

The mean value of $e_{1.23}$ is zero, since those of x_1 and $x_{1.23}$ are both zero. The sum of the products of x_1 and $e_{1.23}$ is

$$\sum x_1 e_{1.23} = \sum x_1(x_1 - x_{1.23}) = \sum x_1^2 - \sum x_{1.23}^2$$
$$= N(\sigma_1^2 - \sigma_{1.23}^2).$$

Also
$$\sum e_{1.23}^2 = \sum (x_1 - x_{1.23})^2 = \sum x_1^2 - \sum x_{1.23}^2$$
$$= N(\sigma_1^2 - \sigma_{1.23}^2).$$

Consequently the coefficient of correlation between x_1 and $e_{1.23}$ is given by

$$R_{1(23)} = \frac{\sigma_1^2 - \sigma_{1.23}^2}{\sigma_1 \sqrt{(\sigma_1^2 - \sigma_{1.23}^2)}} = \frac{1}{\sigma_1}\sqrt{(\sigma_1^2 - \sigma_{1.23}^2)}.$$

This is the required coefficient of multiple correlation. The result may be expressed

$$\sigma_{1.23}^2 = \sigma_1^2(1 - R_{1(23)}^2), \tag{23}$$

which is analogous to (6) and to §41 (46). Comparing (23) and (19) we see that

$$1 - R_{1(23)}^2 = \omega/\omega_{11}, \tag{24}$$

whence
$$R_{1(23)}^2 = 1 - \omega/\omega_{11} = 1 - \omega/(1 - r_{23}^2)$$

$$= \frac{r_{12}^2 + r_{13}^2 - 2r_{12}r_{23}r_{31}}{1 - r_{23}^2}, \tag{25}$$

which expresses the multiple correlation in terms of the total correlations between the pairs of variables.

* Some writers prefer the notation $R_{1.23}$.

K

We may note that $R_{1(23)}$ is never negative. For, from the above argument, $\sum x_1 e_{1.23} = \sum e_{1.23}^2$, which cannot be negative. Further, when $R_{1(23)} = 1$ it follows from (23) that $\sigma_{1.23}^2 = 0$, which requires all the deviations $x_{1.23}$ to be zero, so that x_1 is given accurately by the regression equation. In this case x_1 is a linear function of x_2 and x_3.

The argument is also valid for a distribution of n variables. The multiple correlation $R_{1(23\ldots n)}$ is the correlation between x_1 and its estimate $e_{1.23\ldots n}$ from the regression equation (20); and the argument leads to the corresponding formula

$$\sigma_{1.23\ldots n}^2 = \sigma_1^2(1 - R_{1(23\ldots n)}^2) \tag{23$'$}$$

and

$$1 - R_{1(23\ldots n)}^2 = \omega/\omega_{11}. \tag{24$'$}$$

If the multiple correlation is equal to unity, x_1 is a linear function of $x_2, \ldots, x_n$.

Example 1. Prove the following relations:

$$\sum x_{1.23} e_{1.23} = 0,$$

$$\sum x_1^2 = \sum x_{1.23}^2 + b_{12.3} \sum x_1 x_2 + b_{13.2} \sum x_1 x_3,$$

$$R_{1(23)}^2 = (b_{12.3} \sum x_1 x_2 + b_{13.2} \sum x_1 x_3)/\sum x_1^2.$$

Example 2. Show that, for the example in the preceding section,

$$R_{3(12)} = 0 \cdot 63.$$

113. Partial correlation

Consider next the correlation between the deviations $x_{1.3}$ and $x_{2.3}$. Since $x_{1.3}$ is the deviation of x_1 from its estimate in terms of x_3, we may regard it as that part of the variable x_1 which remains after the influence of x_3 has been eliminated, as far as can be done by a linear equation. A similar interpretation can be given to $x_{2.3}$. Hence the correlation between these deviations may be looked upon as the correlation between x_1 and x_2 after the influence of x_3 has been eliminated. We denote this correlation by $r_{12.3}$, and call it the *partial correlation* between x_1 and x_2 in the trivariate distribution. Having one secondary subscript $r_{12.3}$ is a partial correlation of the first order. It will be remembered that, in calculating the total correlation r_{12}, the values of x_3 are simply ignored. It is therefore a correlation calculated on the assumption that the variables x_1 and x_2 are influenced only by each other, and not by any other variable.

To find the partial correlation $r_{12.3}$ we observe that, by the theorems of § 110,

$$0 = \sum x_{2.3}x_{1.23} = \sum x_{2.3}(x_1 - b_{12.3}x_2 - b_{13.2}x_3)$$
$$= \sum x_1 x_{2.3} - b_{12.3} \sum x_2 x_{2.3} = \sum x_{1.3}x_{2.3} - b_{12.3} \sum x_{2.3}^2,$$

so that
$$b_{12.3} = \sum x_{1.3}x_{2.3} / \sum x_{2.3}^2. \qquad (26)$$

From this result it follows that $b_{12.3}$ is the coefficient of regression of $x_{1.3}$ on $x_{2.3}$. Similarly, $b_{21.3}$ is the coefficient of regression of $x_{2.3}$ on $x_{1.3}$; and the coefficient of correlation between these deviations is therefore given by
$$r_{12.3}^2 = b_{12.3}b_{21.3}. \qquad (27)$$

Since $\sigma_{1.3}$ and $\sigma_{2.3}$ are the standard deviations of $x_{1.3}$ and $x_{2.3}$, the coefficient of partial correlation is connected with the coefficient of partial regression by the usual formulae

$$b_{12.3} = r_{12.3}\sigma_{1.3}/\sigma_{2.3}, \quad b_{21.3} = r_{21.3}\sigma_{2.3}/\sigma_{1.3}. \qquad (28)$$

And it is now clear why the above b's are called coefficients of partial regression. The correlation $r_{12.3}$ has the same sign as $b_{12.3}$, which is the sign of $-\omega_{12}$ by (18). Also, in virtue of (27), $r_{12.3}$ is symmetrical in the primary subscripts. If the values of the b's are substituted from (18) in (27) we obtain

$$r_{12.3}^2 = \frac{\omega_{12}\omega_{21}}{\omega_{11}\omega_{22}} = \frac{\omega_{12}^2}{\omega_{11}\omega_{22}},$$

so that
$$r_{12.3} = \frac{-\omega_{12}}{\sqrt{(\omega_{11}\omega_{22})}} = \frac{r_{12} - r_{13}r_{23}}{\sqrt{\{(1 - r_{13}^2)(1 - r_{23}^2)\}}}. \qquad (29)$$

Similar formulae may be written down for $r_{13.2}$ and $r_{23.1}$. The partial correlations of the first order are thus expressible in terms of the total correlation coefficients.

The partial correlation between x_1 and x_2 in the trivariate distribution is sometimes defined as the correlation between x_1 and x_2 for a constant value of x_3. In general, however, this correlation will depend upon the constant value of x_3 selected. In certain special cases the second definition agrees with the first, and the result is then the same for all constant values of x_3. Necessary and sufficient conditions* for this agreement may be stated:

(a) In the bivariate distribution of x_1 and x_3 (x_2 being ignored), the regression of x_1 on x_3 must be linear, and the standard deviations of all the x_1 arrays ($x_3 =$ const.) must be equal.

(b) In the trivariate distribution the regression of x_1 on x_2 and x_3 must be linear, and the standard deviations of all the x_1 arrays ($x_2 =$ const., $x_3 =$ const.) must be equal.

These conditions are satisfied in the normal trivariate distribution to be considered in § 116.

For a distribution of n variables there are partial correlations of all orders from 1 to $n-2$. Thus if (k) denotes a definite group of secondary subscripts not including i and j, the correlation between the deviations $x_{i.(k)}$ and $x_{j.(k)}$, denoted by $r_{ij.(k)}$, is a partial correlation of order equal to the number m of subscripts in (k). It is connected with the regression coefficients for these deviations, and their standard deviations $\sigma_{i.(k)}$ and $\sigma_{j.(k)}$, by formulae analogous to (27) and (28), namely,

$$r_{ij.(k)}^2 = b_{ij.(k)} b_{ji.(k)}, \tag{30}$$

and

$$b_{ij.(k)} = r_{ij.(k)} \sigma_{i.(k)} / \sigma_{j.(k)}. \tag{31}$$

If (k) includes all the subscripts but i and j, we find as in (29)

$$r_{ij.(k)} = -\omega_{ij} / \sqrt{(\omega_{ii} \omega_{jj})},$$

the symbols on the right being cofactors in the determinant ω of order n.

A partial correlation of order $m+1$ is expressible in terms of those of order m by an equation of the same form as (29), with a set (k) of secondary subscripts added to each coefficient in the formula. Thus

$$r_{ij.h(k)} = \frac{r_{ij.(k)} - r_{ih.(k)} r_{jh.(k)}}{\sqrt{\{(1 - r_{ih.(k)}^2)(1 - r_{jh.(k)}^2)\}}}, \tag{32}$$

where h, i, j are unequal.

Example. Prove that, for the example in § 111,

$$r_{31.2} = 0 \cdot 445, \quad r_{32.1} = 0 \cdot 42.$$

* For a proof see Camp, 1934, 1, pp. 341-2.

114. Reduction formula for the order of a standard deviation

A S.D. of any order may be expressed in terms of a S.D. and a correlation coefficient of lower order. Thus, since

$$\sum x_{1.23}^2 = \sum x_{1.2}(x_1 - b_{12.3}x_2 - b_{13.2}x_3)$$
$$= \sum x_{1.2}^2 - b_{13.2}\sum x_{1.2}x_{3.2},$$

we have, on dividing by N and using (26) with subscripts interchanged,

$$\sigma_{1.23}^2 = \sigma_{1.2}^2(1 - b_{13.2}b_{31.2}) = \sigma_{1.2}^2(1 - r_{13.2}^2), \tag{33}$$

which is of the same form as (6). Since $\sigma_{1.23}^2$ is symmetrical in the secondary subscripts, the subscripts 2 and 3 may be interchanged in the second member of (33). Substituting the value of $\sigma_{1.2}^2$ given by (6) we may also write the result

$$\sigma_{1.23}^2 = \sigma_1^2(1 - r_{12}^2)(1 - r_{13.2}^2). \tag{34}$$

Similar formulae may be written down by cyclic permutation of the subscripts. The equations (33) and (34) show how a S.D. may be expressed in terms of one of lower order, and one or more of the correlation coefficients. Also from (33) it follows that $\sigma_{1.23} \leqslant \sigma_{1.2}$. The estimate of x_1 from x_2 and x_3 is thus in general better than the estimate from x_2 alone, being just as good only if $r_{13.2}$ is zero.

Further, on comparing (34) with (23) we see that

$$1 - R_{1(23)}^2 = (1 - r_{12}^2)(1 - r_{13.2}^2), \tag{35}$$

which is in agreement with the values already found for $R_{1(23)}$ and $r_{13.2}$ in terms of the total correlations. From (35) it is clear that

$$1 - R_{1(23)}^2 \leqslant 1 - r_{12}^2,$$

so that $\qquad\qquad R_{1(23)}^2 \geqslant r_{12}^2.$

Since $R_{1(23)}$ is symmetrical in the subscripts 2 and 3, it follows that this coefficient of multiple correlation is not numerically less than either r_{12} or r_{13}. If then $R_{1(23)}$ is zero, both r_{12} and r_{13} must be zero, and x_1 is then uncorrelated with either x_2 or x_3.

The same argument shows that, in the case of n variables, the equation (33) has the generalization

$$\sigma_{i.j(k)}^2 = \sigma_{i.(k)}^2(1 - r_{ij.(k)}^2), \tag{36}$$

where (k) denotes as usual a group of secondary subscripts. Repeated application of this formula leads to a generalization of (34), namely,

$$\sigma_{1.23\ldots n}^2 = \sigma_1^2(1 - r_{12}^2)(1 - r_{13.2}^2)(1 - r_{14.23}^2) \ldots (1 - r_{1n.23\ldots\overline{n-1}}^2). \quad (37)$$

Comparing this with (23') we see that

$$1 - R_{1(23\ldots n)}^2 = (1 - r_{12}^2)(1 - r_{13.2}^2) \ldots (1 - r_{1n.23\ldots\overline{n-1}}^2). \quad (38)$$

If $R_{1(23\ldots n)} = 0$ each of the correlations in the second member is zero, and also each of the coefficients $r_{12}, r_{13}, \ldots, r_{1n}$. Thus x_1 is then uncorrelated with any of the other variables.

115. Reduction formula for the order of a regression coefficient

To express a coefficient of regression in terms of coefficients of lower order we may proceed as follows. First

$$\sum x_{1.3} x_{2.3} = \sum x_1(x_2 - b_{23}x_3).$$

Then, since $b_{23} = b_{32}\sigma_2^2/\sigma_3^2$, we may write the above equation, in virtue of (26),

$$\sigma_{2.3}^2 b_{12.3} = b_{12}\sigma_2^2 - b_{13}\sigma_3^2 . b_{32}\sigma_2^2/\sigma_3^2,$$

or, by means of (6),

$$\sigma_2^2(1 - r_{23}^2) b_{12.3} = \sigma_2^2(b_{12} - b_{13}b_{32}).$$

Thus we have the required reduction formula

$$b_{12.3} = \frac{b_{12} - b_{13}b_{32}}{1 - b_{23}b_{32}}. \quad (39)$$

This may be expressed in terms of correlations. For, in virtue of (28), it is equivalent to

$$r_{12.3}\frac{\sigma_{1.3}}{\sigma_{2.3}} = \frac{\sigma_1}{\sigma_2}\left(\frac{r_{12} - r_{13}r_{32}}{1 - r_{23}^2}\right). \quad (40)$$

Substituting the values of $\sigma_{1.3}$ and $\sigma_{2.3}$ given by equations of the form (6), we find (29) again.

In the case of n variables, by similar reasoning, we arrive at the generalization of (39),

$$b_{12.3(k)} = \frac{b_{12.(k)} - b_{13.(k)}b_{32.(k)}}{1 - b_{23.(k)}b_{32.(k)}},$$

where (k) has the usual significance.

116. Normal distribution

A generalization of the bivariate normal distribution to the case of three variables may be obtained as follows.* First suppose that the variables x_2 and x_3 are normally distributed about zero means with correlation r_{23}. Then the probability that a pair of values chosen at random will fall in the interval $dx_2 dx_3$ is

$$dP_1 = \frac{dx_2 dx_3}{2\pi\sigma_2\sigma_3\sqrt{\omega_{11}}} \exp\left[-\frac{1}{2\omega_{11}}\left(\frac{x_2^2}{\sigma_2^2}+\frac{x_3^2}{\sigma_3^2}-2r_{23}\frac{x_2 x_3}{\sigma_2\sigma_3}\right)\right],$$

where $\omega_{11} = 1 - r_{23}^2$. Next assume that the regression of x_1 on x_2 and x_3 is linear, and that in each x_1 array the variable is normally distributed with S.D. which is the same for each of these arrays. Then, since the mean of each array is on the plane of regression (7), the S.D. of each array is $\sigma_{1.23}$, given by

$$\sigma_{1.23}^2 = \sigma_1^2 \omega / \omega_{11}. \tag{41}$$

Consequently the probability that x_1, chosen at random in an assigned array, will fall in the interval dx_1 is

$$dP_2 = \frac{dx_1}{\sigma_{1.23}\sqrt{(2\pi)}} \exp\left(-\tfrac{1}{2}x_{1.23}^2/\sigma_{1.23}^2\right)$$

$$= \frac{\sqrt{\omega_{11}}\,dx_1}{\sigma_1\sqrt{(2\pi\omega)}} \exp\left[-\frac{1}{2\omega\omega_{11}}\left(\omega_{11}\frac{x_1}{\sigma_1}+\omega_{12}\frac{x_2}{\sigma_2}+\omega_{13}\frac{x_3}{\sigma_3}\right)^2\right]$$

by (41) and (17). On forming the product $dP_1 dP_2$ we have the probability that a set of values of the three variables, chosen at random, will fall in the interval $dx_1 dx_2 dx_3$, as

$$dP = \frac{dx_1 dx_2 dx_3}{\sigma_1\sigma_2\sigma_3\sqrt{\{(2\pi)^3\omega\}}} \exp\left(-\tfrac{1}{2}\phi\right), \tag{42}$$

where, after a little reduction, it will be found that

$$\phi = \frac{1}{\omega}\left(\omega_{11}\frac{x_1^2}{\sigma_1^2}+\omega_{22}\frac{x_2^2}{\sigma_2^2}+\omega_{33}\frac{x_3^2}{\sigma_3^2}+2\omega_{23}\frac{x_2 x_3}{\sigma_2\sigma_3}+2\omega_{31}\frac{x_3 x_1}{\sigma_3\sigma_1}+2\omega_{12}\frac{x_1 x_2}{\sigma_1\sigma_2}\right)$$
$$\tag{43}$$

The symmetry of (42) and (43) in the subscripts shows that the properties of the three variables are similar. Thus all the regressions

* Cf. Rietz, 1927, 2, pp. 106–7.

are linear. The variance of each array of x_2's in the trivariate distribution is $\omega \sigma_2^2/\omega_{22} = \sigma_{2.13}^2$, and that of each array of x_3's is $\omega \sigma_3^2/\omega_{33} = \sigma_{3.12}^2$.

We may verify the statement made in § 113 that, in the present case, the correlation between x_1 and x_2 for a constant value of x_3 is the partial correlation $r_{12.3}$. For ϕ may be expressed in the form

$$\phi = \frac{1}{\omega}\left[\frac{\omega_{11}}{\sigma_1^2}(x_1 - b_{13}x_3)^2 + \frac{\omega_{22}}{\sigma_2^2}(x_2 - b_{23}x_3)^2 \right.$$
$$\left. + \frac{2\omega_{12}}{\sigma_1\sigma_2}(x_1 - b_{13}x_3)(x_2 - b_{23}x_3)\right] + \frac{x_3^2}{\sigma_3^2}, \quad (44)$$

and, on comparing (42) with § 38 (28), we see that, for a constant value of x_3, x_1 and x_2 are normally distributed about means $b_{13}x_3$ and $b_{23}x_3$ respectively, with correlation

$$-\frac{\omega_{12}}{\sqrt{(\omega_{11}\omega_{22})}} = r_{12.3}$$

as stated. From (42) and (44) it is also clear that the deviations $x_{1.3}$ and $x_{2.3}$ are normally distributed with correlation $r_{12.3}$.

The above results may be extended to the case of n variables.[*]

117. Significance of an observed partial correlation

Fisher has proved that the sampling distribution of a partial correlation coefficient[†] of order k, in samples from a normal population, is of the same form as that of the correlation coefficient for samples from a bivariate normal population, with the sample number N reduced by k. In particular, when the partial correlation in the population is zero, the square of the partial correlation coefficient, r, in samples of N sets of values, is a $\beta_1(\frac{1}{2}, \frac{1}{2}(N-k-2))$ variate. By the argument used in § 89 it then follows that the statistic t defined by

$$t = \frac{r}{\sqrt{(1-r^2)}}\sqrt{(N-k-2)}, \quad (45)$$

conforms to the t distribution for $(N-k-2)$ D.F. We are thus able to test the significance of an observed partial correlation.

* Cf. Yule and Kendall, 1937, 1, pp. 282–4.   † Fisher, 1924, 4.

Example 1. From a random sample of 21 sets of values from a normal population the calculated value of a partial correlation of order three is 0·40. Is this consistent with the assumption that the corresponding partial correlation in the population is zero?

In applying the above test to our assumption we have $\nu = 21 - 3 - 2 = 16$, and

$$t = \frac{0 \cdot 4 \sqrt{(16)}}{\sqrt{(0 \cdot 84)}} = \frac{1 \cdot 6}{0 \cdot 916} = 1 \cdot 74.$$

This value of t is not significant at the 5 % level, so that the observed correlation is not significant of correlation in the population.

From the sampling distribution it follows, as in the case of two variables, that if Fisher's z transformation of § 94 is applied to the above partial correlation of order k, the statistic z is distributed nearly normally with variance $1/(N - k - 3)$. We are thus able to test if an observed partial correlation differs significantly from some assumed value. Similarly, we can test the significance of the difference of the observed partial correlations in two independent samples.

Example 2. From independent samples, of 32 and 23 sets of values, partial correlations of order four are found to be 0·4 and 0·6 respectively. Examine (i) whether the first value is consistent with the assumption of a normal population with a corresponding correlation of 0·7, and (ii) whether the two samples may be regarded as from the same normal population.

(i) From the Table 7 we find that the values $r_1 = 0 \cdot 4$ and $r_0 = 0 \cdot 7$ correspond to $z_1 = 0 \cdot 424$ and $z_0 = 0 \cdot 868$. The deviation of z_1 is therefore 0·444. The S.E. of z_1 is $1/\sqrt{(32 - 4 - 3)} = 0 \cdot 2$. Since the deviation of z_1 is greater than twice the S.E. it is significant. The assumption of a correlation of 0·7 in the population is thus ruled out.

(ii) Corresponding to $r_1 = 0 \cdot 4$ and $r_2 = 0 \cdot 6$ we have $z_1 = 0 \cdot 424$ and $z_2 = 0 \cdot 693$, giving $z_2 - z_1 = 0 \cdot 27$ nearly. The S.E. of the difference of the z's is given by

$$\epsilon = \sqrt{(\tfrac{1}{25} + \tfrac{1}{16})} = \sqrt{(0 \cdot 1025)} = 0 \cdot 32.$$

The difference $z_2 - z_1$, being less than ϵ, is not significant. The samples may thus be regarded as from the same population.

118. Significance of an observed multiple correlation

Consider next the significance of an observed multiple correlation coefficient, R, of the variable x_1 with the p variables $x_2, x_3, ..., x_{p+1}$, calculated from a random sample of N sets of values from a multivariate normal population. Fisher has found the general sampling distribution* of R, and has shown that it depends, not on the whole matrix of correlations between the variables, but simply on the multiple correlation in the population and the sample size, N. In particular, when the multiple correlation coefficient in the population is zero,† R^2 is a $\beta_1(\frac{1}{2}p, \frac{1}{2}(N-p-1))$ variate. In virtue of Theorem VII of § 72 it follows that, in this case, $R^2/(1-R^2)$ is a $\beta_2(\frac{1}{2}p, \frac{1}{2}(N-p-1))$ variate; and therefore, by the argument of § 92, the statistic F defined by

$$F = \frac{R^2}{1-R^2} \frac{N-p-1}{p}, \tag{46}$$

conforms to the F distribution for

$$\nu_1 = p, \quad \nu_2 = N-p-1. \tag{47}$$

To test the hypothesis that the multiple correlation in the population is zero, we have only to determine from the table of F whether the value of this statistic calculated from the sample is significant. This decides the significance of the observed multiple correlation. We may also remark that, since R^2 is a $\beta_1(\frac{1}{2}p, \frac{1}{2}(N-p-1))$ variate in samples from a normal population in which x_1 is uncorrelated with any of the p variables $x_2, x_3, \ldots, x_{p+1}$, the mean value of R^2 is given by

$$E(R^2) = \frac{p}{N-1}.$$

The problem may also be approached from the point of view of analysis of variance and covariance. The estimate $e_{1.(p)}$ of x_1, given by the regression equation of x_1 on the p variables $x_2, \ldots, x_{p+1}$, is of the form

$$e_{1.(p)} = b_{12}x_2 + b_{13}x_3 + \ldots + b_{1(p+1)}x_{p+1}, \tag{48}$$

and this is connected with the corresponding deviation $x_{1.(p)}$ by the equation

$$x_1 = x_{1.(p)} + e_{1.(p)}. \tag{49}$$

---

* Fisher, 1928, 1. See also Wilks, 1932, 1.
† See also Fisher, 1924, 5.

The N values of $x_{1.(p)}$ are connected by the $p+1$ normal equations

$$\sum x_{1.(p)} = 0, \quad \sum x_i x_{1.(p)} = 0 \quad (i = 2, ..., p+1) \tag{50}$$

Squaring (49) and summing over the distribution we see that

$$\sum x_1^2 = \sum x_{1.(p)}^2 + \sum e_{1.(p)}^2, \tag{51}$$

the sum of products vanishing since

$$\sum x_{1.(p)} e_{1.(p)} = \sum x_{1.(p)} (b_{12}x_2 + b_{13}x_3 + ...) = 0,$$

in virtue of (50). The sum in the first member of (51) is equal to NS_1^2, where S_1^2 is the variance of x_1 in the sample; while the first sum on the right has the value $NS_1^2(1 - R^2)$ by (23′). Consequently

$$\sum e_{1.(p)}^2 = NR^2 S_1^2. \tag{52}$$

The equations (50) may be regarded as $p+1$ linear constraints on the N values $x_{1.(p)}$. Then, on the assumption that these deviations are normally distributed, and that R is zero in the population, the sum $\sum x_{1.(p)}^2/\sigma^2$ is distributed like χ^2 with $N-p-1$ D.F., σ^2 being the variance of x_1 in the population. But, by §77, $\sum x_1^2/\sigma^2$ is similarly distributed with $N-1$ D.F.; and therefore, by Theorem II of §81, $\sum e_{1.(p)}^2/\sigma^2$ is distributed like χ^2 with p D.F. Thus the two sums on the right of (51), when divided by $N-p-1$ and p respectively, give independent and unbiased estimates of σ^2. Inserting their values in terms of R^2 we see that the quotient of these estimates

$$F = \frac{R^2}{1-R^2} \frac{N-p-1}{p} \tag{53}$$

conforms to the F distribution for

$$\nu_1 = p, \quad \nu_2 = N-p-1.$$

We thus arrive at the same result as before.* (See also Ex. XII, 13).

Example. In a sample of 25 sets of values from a normal population $R_{1(234)}$ was found to be 0·4. Show that this is not significant of correlation in the population between x_1 and the variables x_2, x_3, x_4.

Here $p = 3$, $\nu_1 = 3$, $\nu_2 = 21$. Hence

$$F = \frac{0·16}{0·84} \times \frac{21}{3} = \frac{4}{3} = 1·33.$$

This is not significant, since the 5 % value of F is about 3·1.

---

* For the distributions of various statistics occurring in this chapter see Bartlett, 1933, 3.

COLLATERAL READING

YULE, 1907, 1.
YULE and KENDALL, 1937, 1, chapter XIV.
RIETZ, 1927, 2, pp. 92–102 and 106–7.
MILLS, 1938, 1, chapter XVI.
RIDER, 1939, 6, §§ 20, 28, 29, 45–8.
KENNEY, 1939, 3, part II, chapter V.
EZEKIEL, 1930, 2, chapters X–XXIII.
GOULDEN, 1939, 2, chapter VIII.
CAMP, 1934, 1, part II, chapter VI.
SNEDECOR, 1938, 3, chapter XIII.
TIPPETT, 1931, 2, chapter XI.
PEARSON, K., 1901, 1 and 1915, 1.
FISHER, 1924, 4 and 1928, 1.
BARTLETT, 1933, 3.
WILKS, 1932, 1.

EXAMPLES XII

1. In a trivariate distribution it is found that

$$\sigma_1 = 3, \quad \sigma_2 = 4, \quad \sigma_3 = 5; \quad r_{23} = 0{\cdot}40, \quad r_{31} = 0{\cdot}60, \quad r_{12} = 0{\cdot}70.$$

Prove that the partial correlations are

$$r_{23.1} = -0{\cdot}035, \quad r_{31.2} = 0{\cdot}49, \quad r_{12.3} = 0{\cdot}63,$$

and that, if the variates are measured from their means, the linear regression equations are

$$x_1 = 0{\cdot}41x_2 + 0{\cdot}23x_3, \quad x_2 = 0{\cdot}96x_1 - 0{\cdot}025x_3, \quad x_3 = 1{\cdot}04x_1 - 0{\cdot}05x_2.$$

Show also that

$$\sigma_{1.23} = 1{\cdot}87, \quad \sigma_{2.31} = 2{\cdot}85, \quad \sigma_{3.12} = 4{\cdot}00,$$
$$R_{1(23)} = 0{\cdot}78, \quad R_{2(31)} = 0{\cdot}70, \quad R_{3(12)} = 0{\cdot}60.$$

2. Show that

$$\sigma_{1.23}\sigma_{2.31}/r_{12.3} = -\omega\sigma_1\sigma_2/\omega_{12},$$

and, more generally,

$$\sigma_{1.23(k)}\sigma_{2.31(k)}/r_{12.3(k)} = -\omega\sigma_1\sigma_2/\omega_{12}.$$

3. From § 114 (38), deduce that

$$1 - r^2_{1n.23\ldots\overline{n-1}} = (1 - R^2_{1(23\ldots n)})/(1 - R^2_{1(23\ldots\overline{n-1})})$$

which shows how a partial correlation coefficient of order $n - 2$ may be expressed in terms of multiple correlation coefficients of orders $n - 1$ and $n - 2$.

4. Prove the identity

$$b_{12.3}b_{23.1}b_{31.2} = r_{12.3}r_{23.1}r_{31.2}.$$

5. Prove the formula

$$b_{12} = (b_{12.3} + b_{13.2}b_{32.1})/(1 - b_{13.2}b_{31.2}),$$

expressing a regression coefficient in terms of coefficients of higher order. Write down the corresponding formula with subscripts 1 and 2 interchanged, multiply together the two equations and take the square root, thus obtaining

$$r_{12} = (r_{12.3} + r_{13.2}r_{23.1})/\sqrt{\{(1 - r^2_{13.2})(1 - r^2_{23.1})\}}$$

expressing a correlation coefficient in terms of coefficients of higher order.

6. Prove the formulae of Ex. 5 with a group (k) of secondary subscripts added to each of the coefficients.

7. Verify the values of ϕ expressed by § 116 (43) and (44).

8. Show that, if $x_3 = ax_1 + bx_2$, the three partial correlations are numerically equal to unity, $r_{13.2}$ having the sign of a, $r_{23.1}$ the sign of b, and $r_{12.3}$ the opposite sign to a/b.

9. For a sample of 30 sets of values from a normal population, $R_{1(23)}$ is found to be $0·5$. Show that this is significant of correlation in the population between x_1 and x_2, x_3. ($F = 4·5, \nu_1 = 2, \nu_2 = 27$.)

10. Show that a partial correlation $r_{12.34} = 0·5$, in a sample of 20 sets of values from a normal population, is significant at the 5 % level.

11. Two independent samples, of 46 and 36 sets of values, give corresponding partial correlations of order three as $0·41$ and $0·66$

respectively. Show that this is not inconsistent with the assumption that the samples are from the same normal population. Show also that an estimate of the partial correlation in the population, by the method of § 96, is 0·53.

12. Show that, corresponding to the first sample in § 117, Ex. 2, the 95 % fiducial limits for the partial correlation in the population are 0·03 and 0·67.

13. Deduce from the argument on p. 259 that, when the multiple correlation in the normal population is zero, $R^2/(1-R^2)$ is a $\beta_2(\frac{1}{2}p, \frac{1}{2}(N-p-1))$ variate and therefore, by § 72, R^2 is a $\beta_1(\frac{1}{2}p, \frac{1}{2}(N-p-1))$ variate.

14. With the notation of § 118 show that, when the multiple correlation in the normal population is zero, the expected value of R in the sample is $\Gamma(\frac{1}{2}(p+1))\,\Gamma(\frac{1}{2}(N-1))/\Gamma(\frac{1}{2}p)\,\Gamma(\frac{1}{2}N)$.

LITERATURE FOR REFERENCE

1875, 1. HELMERT, F. R. Über die Berechnung des wahrscheinlichen Fehlers aus einer endlichen Anzahl wahrer Beobachtungsfehler. *Zeit. für Math. und Physik*, Bd. 20, S. 300–3. See also *Astronomische Nachrichten*, Bd. 85, no. 2039.

1876, 1. HELMERT, F. R. Über die Wahrscheinlichkeit der Potenzsummen der Beobachtungsfehler, und über einige damit in Zusammenhange stehende Fragen. *Zeit. für Math. und Physik*, Bd. 21, S. 192–218. See also *Astronomische Nachrichten*, Bd. 88, no. 2096–7.

1894, 1. PEARSON, KARL. Contributions to the mathematical theory of evolution. *Phil. Trans. Roy. Soc.* A, vol. 185, p. 71.

1896, 1. PEARSON, KARL. Regression, heredity and panmixia. *Phil. Trans. Roy. Soc.* A, vol. 187, p. 253.

1897, 1. YULE, G. U. On the theory of correlation. *Journ. Roy. Stat. Soc.* vol. 60, p. 812.

1898, 1. PEARSON, KARL and FILON, L. N. G. On the probable errors of frequency constants, and on the influence of random selection on variation and correlation. *Phil. Trans. Roy. Soc.* A, vol. 191, p. 229.

2. SHEPPARD, W. F. On the calculation of the most probable values of the frequency constants, from data arranged according to equidistant divisions of the scale. *Proc. Lond. Math. Soc.* vol. 29, p. 353.

1900, 1. PEARSON, KARL. On the criterion that a given system of deviations from the probable, in the case of a correlated system of variables, is such that it can be reasonably supposed to have arisen from random sampling. *Phil. Mag.* vol. 50, pp. 157–75.

1901, 1. PEARSON, KARL. On lines and planes of closest fit to systems of points in space. *Phil. Mag.* Series 6, vol. 2, p. 559.

2. PEARSON, KARL. On the systematic fitting of curves to observations and measurements. *Biometrika*, vol. 1, p. 265 and vol. 2, p. 1.

1902, 1. PEARSON, KARL. On the probable errors of frequency constants. *Biometrika*, vol. 2, pp. 273–81.

1903, 1. THIELE, T. N. *Theory of Observations*. London: Layton.

1905, 1. PEARSON, KARL. *On the General Theory of Skew Correlation and non-linear Regression.* Drapers' Co. Research Memoirs, Biometric Series II.

1907, 1. YULE, G. U. On the theory of correlation for any number of variables, treated by a new system of notation. *Proc. Roy. Soc.* A, vol. 79, pp. 182–93.

1908, 1. 'STUDENT' (Gosset, W. S.). On the probable error of a mean. *Biometrika*, vol. 6, pp. 1–25.

2. 'STUDENT'. On the probable error of a correlation coefficient. *Ibid.* p. 302.

1911, 1. PEARSON, KARL. On the correction necessary for the correlation ratio. *Biometrika*, vol. 8, p. 254.

1912, 1. POINCARÉ, H. *Calcul des Probabilités.* Paris: Gauthier-Villars.

1913, 1. PEARSON, KARL. On the probable errors of frequency constants. *Biometrika*, vol. 9, pp. 1–10.

2. PEARSON, KARL. On the influence of 'broad categories' on correlation. *Ibid.* pp. 116–39.

3. HARRIS, J. A. On the calculation of intra-class and inter-class coefficients of correlation from class-moments, when the number of possible combinations is large. *Ibid.* pp. 446–72.

1915, 1. PEARSON, KARL. On the partial correlation ratio. *Proc. Roy. Soc.* A, vol. 91, p. 492.

2. FISHER, R. A. Frequency distribution of the values of the correlation coefficient in samples from an indefinitely large population. *Biometrika*, vol. 10, pp. 507–21.

1920, 1. PEARSON, KARL. On the fundamental problem of practical statistics. *Biometrika*, vol. 13, p. 1.

2. BOWLEY, A. L. *Elements of Statistics.* Fourth edition. London.

1921, 1. FISHER, R. A. On the 'probable error' of a coefficient of correlation deduced from a small sample. *Metron*, vol. 1, part 4, pp. 3–32.

2. PEARSON, KARL. On a general method of determining the successive terms in a skew regression line. *Biometrika*, vol. 13, p. 296.

1922, 1. FISHER, R. A. On the interpretation of χ^2 from contingency tables, etc. *Journ. Roy. Stat. Soc.* vol. 85, pp. 87–94.

2. FISHER, R. A. The goodness of fit of regression formulae, and the distribution of regression coefficients. *Ibid.* pp. 597–612.

3. PEARSON, KARL. On the χ^2 test of goodness of fit. *Biometrika*, vol. 14, pp. 186 and 418.

4. MINER, J. R. *Tables of $\sqrt{(1-r^2)}$ and $1-r^2$ for use in Partial Correlation and Trigonometry.* Baltimore: Johns Hopkins Press.

5. PEARSON, KARL (ed.). *Tables of the Incomplete Gamma Function.* London: H.M. Stationery Office.

1923, 1. KELLEY, T. L. *Statistical Method.* New York: The MacMillan Co.

1924, 1. RIETZ, H. L. (ed.). *A Handbook of Mathematical Statistics.* Boston: Houghton Mifflin.

2. FISHER, R. A. On the distribution yielding the error functions of several well known statistics. *Proc. Internat. Math. Cong. Toronto*, vol. 2, pp. 805–13.

3. JONES, D. C. *A First Course in Statistics.* Second edition. London: G. Bell and Sons.

1924, 4. FISHER, R. A. The distribution of the partial correlation coefficient. *Metron*, vol. 3, pp. 329–32.

5. FISHER, R. A. The influence of rainfall on the yield of wheat at Rothamsted. *Phil. Trans. Roy. Soc.* B, vol. 213, pp. 89–142.

1925, 1. FISHER, R. A. Applications of Student's distribution. *Metron*, vol. 5, part 3, pp. 90–104.

2. COOLIDGE, J. L. *An Introduction to Mathematical Probability*. Oxford: Clarendon Press.

3. LÉVY, P. *Calcul des Probabilités*. Paris.

1927, 1. BURGESS, R. W. *Introduction to the Mathematics of Statistics*. Boston: Houghton Mifflin.

2. RIETZ, H. L. *Mathematical Statistics*. Carus Mathematical Monographs, no. 3. Chicago: Open Court.

1928, 1. FISHER, R. A. The general sampling distribution of the multiple correlation coefficient. *Proc. Roy. Soc.* A, vol. 121, pp. 654–73.

1929, 1. FISHER, R. A. Moments and product moments of sampling distributions. *Proc. Lond. Math. Soc.* vol. 30, pp. 199–238.

2. WISHART, J. The correlation between product moments of any order in samples from a normal population. *Proc. Roy. Soc. Edin.* vol. 49, pp. 1–13.

1930, 1. RIDER, P. R. A survey of the theory of small samples. *Annals of Math.* vol. 31, pp. 577–628.

2. EZEKIEL, M. *Methods of Correlation Analysis*. New York: Wiley.

3. FISHER, R. A. Inverse probability. *Proc. Camb. Phil. Soc.* vol. 26, pp. 528–35.

1931, 1. IRWIN, J. O. Mathematical theorems involved in analysis of variance. *Journ. Roy. Stat. Soc.* vol. 94, pp. 284–300.

2. TIPPETT, L. H. C. *The Methods of Statistics*. London: Williams and Norgate.

3. NEYMAN, J. and PEARSON, E. S. On the problem of *k* samples. *Bull. Acad. Polonaise des Sciences et des Lettres*, Series A, p. 460.

4. THIELE, T. N. *Theory of Observations*. English version, reprinted in *Annals of Math. Stat.* vol. 2, p. 165.

1932, 1. WILKS, S. S. On the sampling distribution of the multiple correlation coefficient. *Annals of Math. Stat.* vol. 3, pp. 196–203.

1933, 1. FISHER, R. A. The concepts of inverse probability and fiducial probability referring to unknown parameters. *Proc. Roy. Soc.* A, vol. 139, pp. 343–8.

2. AITKEN, A. C. On the graduation of data by the orthogonal polynomials of least squares. *Proc. Roy. Soc. Edin.* vol. 53, pp. 54–78.

3. BARTLETT, M. S. On the theory of statistical regression. *Ibid.* pp. 260–83.

1934, 1. CAMP, B. H. *The Mathematical Part of Elementary Statistics.*
New York: Heath.

2. SNEDECOR, G. W. *Calculation and Interpretation of Analysis of Variance and Covariance.* Iowa: Collegiate Press.

3. WISHART, J. Statistics in agricultural research. *Supp. to Journ. Roy. Stat. Soc.* vol. 1, pp. 26–51.

4. IRWIN, J. O. Independence of the constituent items in the analysis of variance. *Ibid.* pp. 236–51.

5. YATES, F. The analysis of multiple classifications with unequal numbers in different classes. *Journ. Amer. Stat. Assoc.* vol. 29, p. 56.

6. COCHRAN, W. G. The distribution of quadratic forms in a normal system, etc. *Proc. Camb. Phil. Soc.* vol. 30, pp. 178–91.

1935, 1. FISHER, R. A. The mathematical distributions used in the common tests of significance. *Econometrica,* vol. 3, pp. 353–65.

2. FISHER, R. A. The fiducial argument in statistical inference. *Annals of Eugenics,* vol. 6, pp. 391–8.

1936, 1. LEVY, H. and ROTH, L. *Elements of Probability.* Oxford: Clarendon Press.

2. WISHART, J. Tests of significance in analysis of covariance. *Supp. to Journ. Roy. Stat. Soc.* vol. 3, pp. 79–82.

3. WISHART, J. and SANDERS, H. G. *Principles and Practice of Field Experimentation.* Empire Cotton-Growing Corporation.

1937, 1. YULE, G. U. and KENDALL, M. G. *Introduction to the Theory of Statistics.* Eleventh edition. London: Griffin and Co.

2. USPENSKY, J. V. *Introduction to Mathematical Probability.* New York: McGraw-Hill.

3. PITMAN, E. J. G. Significance tests which may be applied to samples from any population. *Supp. to Journ. Roy. Stat. Soc.* vol. 4, pp. 119–30 and 225–32.

4. YATES, F. *The Design and Analysis of Factorial Experiments.* Imperial Bureau of Soil Science, Harpenden, Herts.

5. RIETZ, H. L. Some topics in sampling theory. *Bull. Amer. Math. Soc.* vol. 43, pp. 209–30.

6. CRAMÉR, H. *Random Variables and Probability Distributions.* Cambridge: University Press.

7. CORNISH, E. A. and FISHER, R. A. Moments and cumulants in the specification of distributions. *Revue de l'Institut International de Statistique,* vol. 4, pp. 1–14.

8. PITMAN, E. J. G. The 'closest' estimates of statistical parameters. *Proc. Camb. Phil. Soc.* vol. 33, pp. 212–22.

1938, 1. MILLS, F. C. *Statistical Methods.* Revised edition. London: Isaac Pitman.

2. FISHER, R. A. *Statistical Methods for Research Workers.* Seventh edition. Edinburgh: Oliver and Boyd.

3. SNEDECOR, G. W. *Statistical Methods.* Revised edition. Iowa: Collegiate Press.

1938, 4. YATES, F. Orthogonal functions and tests of significance in the analysis of variance. *Supp. to Journ. Roy. Stat. Soc.* vol. 5, pp. 177–80.

5. KENDALL, M. G. The conditions under which Sheppard's corrections are valid. *Journ. Roy. Stat. Soc.* vol. 101, pp. 592–605.

6. RIETZ, H. L. On a recent advance in statistical inference. *Amer. Math. Monthly*, vol. 45, pp. 149–58.

7. FISHER, R. A. and YATES, F. *Statistical Tables.* Edinburgh: Oliver and Boyd.

1939, 1. AITKEN, A. C. *Statistical Mathematics.* Edinburgh: Oliver and Boyd.

2. GOULDEN, C. H. *Methods of Statistical Analysis.* New York: Wiley and Sons.

3. KENNEY, J. F. *Mathematics of Statistics.* New York: Van Nostrand.

4. KURTZ, A. K. and EDGERTON, H. A. *Statistical Dictionary of Terms and Symbols.* New York: Wiley and Sons.

5. JEFFREYS, H. *Theory of Probability.* Oxford: Clarendon Press.

6. RIDER, P. R. *An Introduction to Modern Statistical Methods.* New York: Wiley and Sons.

7. SMITH, J. H. *Tests of Significance. What they mean and how to use them.* Chicago: University Press.

1940, 1. PLUMMER, H. C. *Probability and Frequency.* London: MacMillan.

2. PETERS, C. C. and VAN VOORHIS, W. R. *Statistical Procedures and their Mathematical Bases.* New York: McGraw-Hill.

3. SAWKINS, D. T. Elementary presentation of the frequency distributions of certain statistical populations associated with the normal population. *Journ. and Proc. Roy. Soc. N.S.W.* vol. 74, pp. 209–39.

4. KENDALL, M. G. Note on the distribution of quantiles for large samples. *Supp. to Journ. Roy. Stat. Soc.* vol. 7, pp. 83–5.

5. WISHART, J. *Field trials. Their lay-out and statistical analysis.* Imperial Bureau of Plant Breeding and Genetics, School of Agriculture, Cambridge.

6. PITMAN, E. J. G. Tests of hypotheses concerning location and scale parameters. *Biometrika*, vol. 31, pp. 200–15.

7. HSU, P. L. On generalized Analysis of Variance. *Ibid.* pp. 221–37.

8. KENDALL, M. G. Some properties of k-statistics. *Annals Eugenics*, vol. 10, pp. 106–11, 215–22 and 392–402.

1941, 1. SAWKINS, D. T. Remarks on goodness of fit of hypotheses and on Pearson's χ^2-test. *Journ. and Proc. Roy. Soc. N.S.W.* vol. 75, pp. 85–95.

2. KENDALL, M. G. A theory of Randomness. *Biometrika*, vol. 32, pp. 1–15.

3. BARTLETT, M. S. The statistical significance of canonical correlations. *Ibid.* pp. 29–37.

4. HSU, P. L. Analysis of Variance from the power function standpoint. *Ibid.* pp. 62–9.

1941, 5. NEYMAN, J. Fiducial argument and the theory of confidence intervals. *Ibid.* pp. 128–50.
6. HOTELLING, H. Experimental determination of the maximum of a function. *Ann. Math. Stat.* vol. 12, pp. 20–45.
7. WILKS, S. S. Determination of sample sizes for setting tolerance limits. *Ibid.* pp. 91–96.
8. WALD, A. and BROOKNER, R. J. On the distribution of Wilks' statistic. *Ibid.* pp. 137–52.

1942, 1. HALDANE, J. B. S. The mode and median of a nearly normal distribution with given cumulants. *Biometrika*, vol. 32, pp. 294–9.
2. HARTLEY, H. O. The range in random samples. *Ibid.* pp. 334–48.
3. GEARY, R. C. Inherent relations between random variables. *Proc. Royal Irish Acad.* vol. 47, pp. 63–76.
4. NEYMAN, J. Basic ideas of the theory of testing statistical hypotheses. *Journ. Roy. Stat. Soc.* vol. 105, pp. 292–327.
5. AITKEN, A. C. and SILVERSTONE, H. On the estimation of statistical parameters. *Proc. Roy. Soc. Edinb.* A, vol. 61, pp. 186–94.
6. ANDERSON, R. L. Distribution of the serial correlation coefficient. *Ann. Math. Stat.* vol. 13, pp. 1–13.
7. WALD, A. Setting of tolerance limits when the sample is large. *Ibid.* pp. 389–99.
8. WILKS, S. S. Statistical prediction with special reference to the problem of tolerance limits. *Ibid.* pp. 400–9.

1943, 1. WILKS, S. S. *Mathematical Statistics.* Princeton University Press.
2. KENDALL, M. G. *The Advanced Theory of Statistics.* Vol. 1. London: Griffin and Co.
3. VAJDA, S. The algebraic analysis of contingency tables. *Journ. Roy. Stat. Soc.* vol. 106, pp. 333–42.
4. WALD, A. An extension of Wilks' method for setting tolerance limits. *Ann. Math. Stat.* vol. 14, pp. 45–55.
5. CURTISS, J. H. On transformations used in the Analysis of Variance. *Ibid.* pp. 107–22.
6. MISES, R. v. On the problem of testing hypotheses. *Ibid.* pp. 238–52.

1944, 1. SAWKINS, D. T. Simple regression and correlation. *Journ. and Proc. Roy. Soc. N.S.W.* vol. 77, pp. 85–95.
2. KENDALL, M. G. On autoregressive time series. *Biometrika*, vol. 33, pp. 105–22.
3. TIPPETT, L. H. C. The control of industrial processes subject to trends in quality. *Ibid.* pp. 163–72.
4. HARTLEY, H. O. Studentization. *Ibid.* pp. 173–80.
5. WALD, A. On a statistical problem arising in the classification of an individual into one of two groups. *Ann. Math. Stat.* vol. 15, pp. 145–62.
6. WALD, A. and WOLFOWITZ, J. Statistical tests based on permutations of observations. *Ibid.* pp. 358–72.
7. GUMBEL, E. J. Ranges and midranges. *Ibid.* pp. 414–22.

Literature for Reference 269

1945, 1. KENDALL, M. G. The treatment of ties in ranking problems. *Biometrika*, vol. 33, pp. 239–51.
  2. PEARSON, E. S., GODWIN, H. J. and HARTLEY, H. O. The probability integral of the mean deviation. *Ibid.* pp. 252–65.
  3. AITKEN, A. C. On linear approximation by least squares. *Proc. Roy. Soc. Edinb.* A, vol. 62, pp. 138–46.
  4. HSU, P. L. The approximate distributions of the mean and variance of a sample of independent variables. *Ann. Math. Stat.* vol. 16, pp. 1–29.
  5. WALD, A. Sequential tests of statistical hypotheses. *Ibid.* pp. 117–86.
  6. FELLER, W. On the normal approximation to the binomial distribution. *Ibid.* pp. 319–29.
  7. BRONOWSKI, J. and NEYMAN, J. The variance of the measure of a two-dimensional random set. *Ibid.* pp. 330–41.

1946, 1. CRAMÉR, H. *Mathematical Methods of Statistics.* Princeton University Press.
  2. KENDALL, M. G. *The Advanced Theory of Statistics.* Vol. 2. London: Griffin and Co.
  3. YATES, F. A review of recent statistical developments in sampling and sampling surveys. *Journ. Roy. Stat. Soc.* vol. 109, pp. 12–30.
  4. BARTLETT, M. S. On the theoretical specification of sampling properties of autocorrelated time series. *Supp. to Journ. Roy. Stat. Soc.* vol. 8, pp. 27–41.
  5. BROWN, G. W. and TUKEY, J. W. Some distributions of sample means. *Ann. Math. Stat.* vol. 17, pp. 1–12.
  6. GIRSHICK, M. A. Contributions to the theory of sequential analysis. *Ibid.* pp. 123–43 and 282–98.
  7. WALD, A. and WOLFOWITZ, J. Tolerance limits for a normal distribution. *Ibid.* pp. 208–15.
  8. WILKS, S. S. Sample criteria for testing equality of means, of variances and of covariances in a normal multivariate distribution. *Ibid.* pp. 257–81.

1947, 1. WALD, A. *Sequential Analysis.* New York: Wiley and Sons.
  2. HOEL, P. G. *Introduction to Mathematical Statistics.* New York: Wiley and Sons.
  3. BARTLETT, M. S. Multivariate Analysis. *Supp. to Journ. Roy. Stat. Soc.* vol. 9, pp. 176–90.
  4. PEARSON, E. S. The choice of statistical tests. *Biometrika*, vol. 34, pp. 139–67.
  5. GEARY, R. C. Testing for normality. *Ibid.* pp. 209–42.
  6. HOTELLING, H. A generalized T measure of multivariate dispersion. *Ann. Math. Stat.* vol. 18, p. 298.

1948, 1. PITMAN, E. J. G. Lecture notes on *Non-parametric Statistical Inference.* University of North Carolina, Institute of Statistics. (Mimeographed.)

270 *Literature for Reference*

1948, 2. WOLD, H. O. A. *Random Normal Deviates.* Tracts for Computers, No. 24. Cambridge University Press.

3. KENDALL, M. G. *Rank Correlation Methods.* London: Griffin and Co.

4. HARTLEY, H. O. The estimation of non-linear parameters by 'internal least squares'. *Biometrika*, vol. 35, pp. 32–45.

5. RADHAKRISHNA RAO, C. Tests of significance in multivariate analysis. *Ibid.* pp. 58–79.

6. STEVENS, W. L. Control of gauging. *Journ. Roy. Stat. Soc. B,* vol. 10, pp. 54–98.

7. COCHRAN, W. G. and BLISS, C. I. Discriminant functions with covariance. *Ann. Math. Stat.* vol. 19, pp. 151–76.

8. WALD, A. and WOLFOWITZ, J. Optimum character of the sequential probability ratio test. *Ibid.* pp. 326–39.

1949, 1. DAVID, F. N. *Probability Theory for Statistical Methods.* Cambridge University Press.

2. NEYMAN, J. (Ed.). *Proceedings of the Berkeley Symposium on Mathematical Statistics and Probability.* University of California Press.

3. BARNARD, G. A. Statistical Inference. *Journ. Roy. Stat. Soc. B,* vol. 11, pp. 115–39.

4. MOYAL, J. E., BARTLETT, M. S. and KENDALL, D. G. Symposium on stochastic processes. *Ibid.* pp. 150–264.

5. WISHART, J. Cumulants of multivariate multinomial distributions. *Biometrika*, vol. 36, pp. 47–58.

6. LANCASTER, H. O. The derivation and partition of χ^2 in certain discrete distributions. *Ibid.* pp. 117–29 and 370–82.

7. WALSH, J. E. On significance tests for the median. *Ann. Math. Stat.* vol. 20, pp. 64–81.

8. WALD, A. Statistical decision functions. *Ibid.* pp. 165–205.

1950, 1. FELLER, W. *An Introduction to Probability Theory and its Applications.* New York: Wiley and Sons.

2. FISHER, R. A. *Contributions to Mathematical Statistics.* New York: Wiley and Sons.

3. WALD, A. *Statistical Decision Functions.* New York: Wiley and Sons.

4. DEMING, W. E. *Some Theory of Sampling.* New York: Wiley and Sons.

5. MOOD, A. McF. *Introduction to the Theory of Statistics.* New York: McGraw-Hill.

6. PEARSON, E. S. On questions raised by the combination of tests based on discontinuous distributions. *Biometrika*, vol. 37, pp. 383–98.

7. MORAN, P. A. P. Recent developments in ranking theory. *Journ. Roy. Stat. Soc. B,* vol. 12, pp. 153–62. See also pp. 292–5.

8. LEHMANN, E. L. Some principles of the theory of testing hypotheses. *Ann. Math. Stat.* vol. 21, pp. 1–26.

1951, 1. DIXON, W. J. and MASSEY, F. J. *Introduction to Statistical Analysis.* New York: McGraw-Hill.

2. RIDER, P. R. The distribution of the range in samples from a discrete rectangular population. *Journ. Amer. Stat. Ass.* vol. 46, pp. 375–8 and 502–7.

3. BARNARD, G. A. The theory of information. *Journ. Roy. Stat. Soc. B,* vol. 13, pp. 46–64.

4. RAMAKRISHNAN, A. Some simple stochastic processes. *Ibid.* pp. 131–40.

5. KENDALL, M. G. Regression, structure and functional relationship. I. *Biometrika,* vol. 38, pp. 11–25.

6. KEEPING, E. S. A significance test for exponential regression. *Ann. Math. Stat.* vol. 22, pp. 180–98.

1952, 1. DUNCAN, A. J. *Quality Control and Industrial Statistics.* Chicago: Richard D. Irwin.

2. WILLIAMS, E. J. Some exact tests in multivariate analysis. *Biometrika,* vol. 39, pp. 17–31. See also pp. 65–81.

3. KENDALL, M. G. Regression, structure and functional relationship. II. *Ibid.* pp. 96–108.

4. TOCHER, K. D. On the concurrence of a set of regression lines. *Ibid.* pp. 109–17.

5. BARNARD, G. A. The frequency justification of certain sequential tests. *Ibid.* pp. 144–50.

6. COX, D. R. Estimation by double sampling. *Ibid.* pp. 217–27.

7. CHERNOFF, H. and SCHEFFÉ, H. A generalization of the Neyman-Pearson fundamental lemma. *Ann. Math. Stat.* vol.23, pp. 213–25.

8. COCHRAN, W. G. The χ^2-test of goodness of fit. *Ibid.* pp. 315–45.

1953, 1. COCHRAN, W. G. *Sampling Techniques.* New York: Wiley and Sons.

2. DOOB, J. L. *Stochastic Processes.* New York: Wiley and Sons.

3. RIDER, P. R. The distribution of the product of ranges in samples from a rectangular population. *Journ. Amer. Stat. Ass.* vol. 48, pp. 546–9. See also pp. 826–30.

4. BARTLETT, M. S. Approximate confidence intervals. *Biometrika,* vol. 40, pp. 12–19 and 306–17.

5. ANSCOMBE, F. J. Sequential Estimation. *Journ. Roy. Stat. Soc. B,* vol. 15, pp. 1–21.

6. LINDLEY, D. V. Statistical Inference. *Ibid.* pp. 30–65.

7. HOTELLING, H. New light on the correlation coefficient and its transforms. *Ibid.* pp. 193–225.

8. McMILLAN, B. The basic theorems of information theory. *Ann. Math. Stat.* vol. 24, pp. 196–219.

1954, 1. SAVAGE, L. J. *The Foundations of Statistics.* New York: Wiley and Sons.

2. FISHER, R. A. The Analysis of Variance with various binomial transformations. *Biometrics,* vol. 10, pp. 130–9.

272 *Literature for Reference*

1954, 3. DUNNETT, C. W. and SOBEL, M. A bivariate generalization of Student's *t*-distribution. *Biometrika*, vol. 41, pp. 153–69.

4. BARNARD, G. A. Simplified decision functions. *Ibid.* pp. 241–51.

5. WHITTLE, P. On stationary processes in the plane. *Ibid.* pp. 434–49.

6. JAMES, A. T. Normal multivariate analysis and the orthogonal group. *Ann. Math. Stat.* vol. 25, pp. 40–75.

7. HARTLEY, H. O. and DAVID, H. A. Universal bounds for mean range and extreme observation. *Ibid.* pp. 85–99.

8. RUTHERFORD, R. S. G. On a contagious distribution. *Ibid.* pp. 703–13.

1955, 1. BARTLETT, M. S. *An Introduction to Stochastic Processes.* Cambridge University Press.

2. HANNAN, E. J. Exact tests for serial correlation. *Biometrika*, vol. 42, pp. 133–42. See also pp. 316–26.

3. WATSON, G. S. Serial correlation in regression analysis. I. *Ibid.* pp. 327–41.

4. YATES, F. The use of transformations and maximum likelihood in the analysis of quantal experiments involving two treatments. *Ibid.* pp. 382–403.

5. JOWETT, G. H. The comparison of means of sets of observations from sections of independent stochastic series. *Journ. Roy. Stat. Soc.* B, vol. 17, pp. 208–27.

6. LE CAM, L. An extension of Wald's theory of statistical decision functions. *Ann. Math. Stat.* vol. 26, pp. 69–81.

7. KAC, M., KIEFER, J. and WOLFOWITZ, J. On tests of normality and other tests of goodness of fit. *Ibid.* pp. 189–211.

1956, 1. BARTHOLOMEW, D. J. A sequential test of randomness for events occurring in time or space. *Biometrika*, vol. 43, pp. 64–78.

2. DANIELS, H. E. The approximate distribution of serial correlation coefficients. *Ibid.* pp. 169–85.

3. QUENOUILLE, M. H. Notes on bias in estimation. *Ibid.* pp. 353–60.

4. WATSON, G. S. and HANNAN, E. J. Serial correlation in regression analysis. II. *Ibid.* pp. 436–48.

5. BARTON, D. E. and DAVID, F. N. Some notes on ordered random intervals. *Journ. Roy. Stat. Soc.* B, vol. 18, pp. 79–94.

6. MALLOWS, C. L. Generalizations of Tchebycheff's inequalities. *Ibid.* pp. 140–68.

7. KEMP, C. D. and KEMP, A. W. Generalized hypergeometric distributions. *Ibid.* pp. 202–11.

8. CHERNOFF, H. Large sample theory. *Ann. Math. Stat.* vol. 27, pp. 1–22.

INDEX

The numbers refer to the pages